G. Geoffrey Vining

Statistics for the 21st Century: Methodologies for Applications of the Future, edited by C. R. Rao and Gábor J. Székely

Probability and Statistical Inference, Nitis Mukhopadhyay

Handbook of Stochastic Analysis and Applications, edited by D. Kannan and V. Lakshmikantham

Testing for Normality, Henry C. Thode, Jr.

Handbook of Applied Econometrics and Statistical Inference, edited by Aman Ullah, Alan T. K. Wan, and Anoop Chaturvedi

Visualizing Statistical Models and Concepts, R. W. Farebrother and Michaël Schyns

Financial and Actuarial Statistics: A Introduction, Dale S. Borowiak

Nonparametric Statistical Inference, Fourth Edition, Revised and Expanded, Jean Dickinson Gibbons and Subhabrata Chakraborti

Computer-Aided Econometrics, edited by David E.A. Giles

The EM Algorithm and Related Statistical Models, edited by Michiko Watanabe and Kazunori Yamaguchi

Multivariate Statistical Analysis, Second Edition, Revised and Expanded, Narayan C. Giri

Computational Methods in Statistics and Econometrics, Hisashi Tanizaki

Applied Sequential Methodologies: Real-World Examples with Data Analysis, edited by Nitis Mukhopadhyay, Sujay Datta, and Saibal Chattopadhyay

Handbook of Beta Distribution and Its Applications, edited by Arjun K. Gupta and Saralees Nadarajah

Item Response Theory: Parameter Estimation Techniques, Second Edition, edited by Frank B. Baker and Seock-Ho Kim

Statistical Methods in Computer Security, edited by William W. S. Chen

Elementary Statistical Quality Control, Second Edition, John T. Burr

A Kalman Filter Primer

R. L. Eubank

Department of Mathematics and Statistics
Arizona State University
Tempe, AZ, USA

CRC Press
Taylor & Francis Group
Boca Raton London New York

CRC Press is an imprint of the
Taylor & Francis Group, an **informa** business

A CHAPMAN & HALL BOOK

CRC Press
Taylor & Francis Group
6000 Broken Sound Parkway NW, Suite 300
Boca Raton, FL 33487-2742

First issued in paperback 2019

© 2006 by Taylor & Francis Group, LLC
CRC Press is an imprint of Taylor & Francis Group, an Informa business

No claim to original U.S. Government works

ISBN-13: 978-0-8247-2365-1 (hbk)
ISBN-13: 978-0-367-39169-0 (pbk)

Library of Congress Cataloging-in-Publication Data

Catalog record is available from the Library of Congress

**Visit the Taylor & Francis Web site at
http://www.taylorandfrancis.com**

**and the CRC Press Web site at
http://www.crcpress.com**

In loving memory of Boomer and Tiny

Preface

My first exposure to the Kalman filter was through a project for a graduate course taught by Jim Matis at Texas A&M University in 1978. Most of the literature on Kalman filtering was written by engineers at that time and I can remember puzzling about the attempts at motivation through circuit diagrams, etc., that I would find in this work. The intent of such artwork was unfortunately lost on an Agricultural Economist with statistical ambitions like myself. Perhaps most perplexing was the initializing state vector and the absence of details concerning its choice left me baffled.

After (more or less) completing the project, I put the Kalman filter on the back burner for over 20 years. I finally returned to it in the summer of 1999 when I made a serious attempt at understanding Craig Ansley's and Robert Kohn's method of using the Kalman filter to compute a smoothing spline.

The Kohn/Ansley work on filtering is very much from a Bayesian and, more to the point, normal theory perspective. As someone with essentially a frequentist background I wanted to see things my own way which, in this setting, meant in terms of regression (or Fourier) analysis and the Cholesky decomposition. Not surprisingly, this is the viewpoint that I have used throughout the book (with the exception of Chapter 7).

My goal in writing this book was to produce the type of text I wished had been available to me in the summer of 1999. What I wanted was a self-contained, "no frills," mathematically rigorous derivation of all the basic Kalman filter recursions from first principles. While there were many books that had some of these characteristics, I found none that had them all.

The "no frills" part of my "wish book" criterion manifests in several aspects of my approach to writing the present text. First, I consider very few examples. In addition, the state-space model I have focused on throughout most of the book is a pared down version of more general models (see, e.g., Chapter 8) that one often finds in the literature. In my opinion, this simplification produces substantial notational savings with very little conceptual downside. There is no doubt that complex notation can create a learning barrier and the notation involved in the Kalman filter can be quite formidable. Consequently, I have focused on a simpler model with the hopes that this will make the reader's initial contact with the Kalman filter a bit less overwhelming.

To produce a self-contained treatment I have treated only one topic: namely, the Kalman filter. In addition, the writing style is intended to be genre generic and all but a few of the tools I use are developed directly in the text. Exceptions to the latter point include results from standard mathematical statistics, regression and multivariate analysis. This is not a book on time series analysis or even state-space models, for that matter. Readers who are seeking that type of material should explore other avenues such as Durbin and Koopman (2001).

Finally, the "mathematically rigorous" aspect of the book is aimed at giving the reader a fundamental understanding of how the Kalman filter works. For that reason essentially every major result is proved in detail. The techniques employed in the proofs are fundamental to the area and hence can be adapted to solve other related problems of interest. My goal in studying the Kalman

filter was to understand its inner workings at a deep e-
nough level that I would be comfortable using it in my
own research and, perhaps the real acid test, become ca-
pable of writing code to implement it in practice. I have
written this text for those who desire a similar level of
immersion in the topic.

Both the nondiffuse and diffuse Kalman filters readily lend themselves to code development and I have tried
to emphasize this by including pseudo-code algorithmic
summaries at various points throughout the text. Imple-
mentations of these algorithms in Java can be download-
ed from math.asu.edu/~eubank.

The book is organized as follows. First, in Chapter 1,
I lay out the basic prediction problem for signal-plus-
noise models, which include state-space models as a special case. The Gramm-Schmidt algorithm and Cholesky
decomposition are derived in this setting and then spe-
cialized to state-space models, which have the additional problem of state vector estimation, at the end of the
chapter.

Chapter 2 gives a complete, inherently recursive, characterization of the covariance between the state and innovation vectors. This is the basic tool employed for deriving the forward and backward (or smoothing) Kalman
recursions in Chapters 4 and 5 as well as the algorithms
for computing Cholesky factorizations and inverse matrices in Chapter 3. The key point is that the Kalman filter is basically a modified Cholesky algorithm that uses
the extra structure obtained from state-space models to
compute predictions orders of magnitude more efficiently than is generally possible for their signal-plus-noise
parent.

Chapter 6 deals with the choice of the initial state vector. It turns out that I had reason to worry about this
point in 1978. The issue was not resolved satisfactorily
until the mid 1980's.

Chapter 7 examines the special case of normal state-
space models. In that setting, the primary focus is on

evaluating the likelihood under both nondiffuse and diffuse prior distributions for the initial state. The study of state-space models concludes in Chapter 8 where the results from Chapters 4–7 are extended to a model formulation that is sufficiently general to account for most situations that might arise in practice.

There are few books that are truly solo efforts and this particular one is no exception. Most of my initial study of Kalman filtering was conducted jointly with my friend and erstwhile co-author Suojin Wang. This book would not have been possible without the insights and skills I have gained through collaboration with Suojin.

There are many fewer errors in this text as a result of the careful proofreading conducted by my friends (and graduate students) Yolanda Muñoz and Hyejin Shin. Help with various formatting and LATEX issues provided by Jessika Vakili and Nishith Arora is also gratefully acknowledged. Kevin Sequeira resurrected this project from near death and has provided both encouragement and guidance on creating the type of end product I originally had in mind when I wrote the first partial draft. The hand written version of this initial draft was skillfully converted to LATEX by Elaine Washington. A helpful discussion with Andrew Bremner about continued fractions substantially improved my treatment of the case where H, F, W and Q are not time dependent.

Finally, I have been fortunate to have my best friends Lisa, Daisy and Shadow to keep me company and in "good" spirits throughout yet another protracted writing effort. Their love and understanding humbles me and allows me to assign a personal meaning to the word "blessing."

RANDALL L. EUBANK

Contents

1

Signal-Plus-Noise Models

1.1 Introduction

Both the theory and practice of statistics frequently involves consideration of stochastic models evolving with respect to a discrete, time-like, index variable. This book is concerned with an important special class of such stochastic processes that can be described using *state-space models*. In this and subsequent chapters we will develop a mathematical framework that can be used to understand the computational algorithms that are widely used in conducting statistical inference about these processes.

State-space models are examples of signal-plus-noise models and, accordingly, we will begin the development in Section 1.2 in the more general signal-plus-noise environment. This provides us with a broader view of prediction problems while also allowing us to appreciate the benefits that can be realized when a state-space framework is appropriate. In particular, we will see that for state-space models it is possible to develop algorithms for computing predictions, parameter estimators, etc., that are orders of magnitude faster than for the general signal-plus-noise setting. These algorithms are usually referred to under the titular umbrella of the *Kalman filter* as a result of the pioneering work in Kalman (1960) and Kalman and Bucy (1961).

1.2   The prediction problem

The standard statistical prediction problem involves using the *observed* values of one set of random variables (i.e.,
the predictors) to estimate or, more correctly, predict
the *unobserved* values of another set of random variables
(i.e., the predictands). The predictors and predictand-
s must be dependent (in a probabilistic sense) in order
for there to be any hope of accomplishing such a task.
The amount of knowledge one has about this dependence
then determines the specifics of how the predictors are
utilized in constructing a prediction formula or prescription that translate predictors into actual predictions for
the predictands. Given a complete specification of the
joint distribution of the predictors and predictands, it
may be possible to obtain predictions that are optimal in
the sense of some performance criterion. However, there
is no universally acknowledged "best" performance criterion and cases where one has complete knowledge of
joint probability distributions are virtually nonexistent
in practice.

There are various paradigms for carrying out predictions when one possesses only partial knowledge about
dependence structure. Least-squares is one such method that requires only second order (i.e., covariance) information and provides "optimal" predictions in various
respects when the random variables in question are jointly normally distributed. There are other reasons to employ least-squares; not the least of these is the property
that least-squares prediction prescriptions are typically
mathematically and computationally tractable. This is
particularly true when the prescription is taken to be
linear in the predictors. Accordingly, we will restrict our
attention to linear least-squares prediction in all that
follows.

To formulate the prediction problem we will think of

the random variables in question as forming a *time se-ries* or *stochastic process*. By stochastic process we mean a collection of random variables that are indexed by a set T. More precisely, suppose that for each $t \in T$ there is an associated random variable $y(t)$. Then, this gives us the stochastic process (or merely process if the context is clear) $\{y(t) : t \in T\}$. When discussing a process $\{y(t) : t \in T\}$ we will, on occasion, refer to the entire collection of random variables as being the y process or simply as $y(\cdot)$. When T corresponds to some subset of the integers or the real line $y(\cdot)$, is sometimes called a time series.

On occasion we will use the phrase "normal" to de-scribe a stochastic process or time series $\{y(t) : t \in T\}$. This indicates, among other things, that the random variables $y(t)$ are normally distributed for each t in T. But, there is actually much more to being "normal" than this. In fact, for a process to be *normal* it must be true that the collection of random variables $y(t_1), \ldots, y(t_k)$ have a joint normal distribution for any set of indices $t_1, \ldots, t_k$ in T and any finite integer k. An explanation and discussion of the consequences of this definition for "normality" can be found in, e.g., Doob (1953, Section II.3).

Our development will focus entirely on time series with the discrete time index $T = \{1, \ldots, n\}$ although simi-lar treatments are also possible in the continuous time case. (See, e.g., Chapter 4 of Gelb 1974.) In addition, we will deal with the somewhat more general case where the y process is vector valued. That is, for each $t \in T$ we observe p random variables $y_1(t), \ldots, y_p(t)$ that are concatenated to form the random column vector $y(t) = (y_1(t), \ldots, y_p(t))^T$. By working with vector, rather than scalar, processes we gain considerable generality while incurring only a minimal level of additional notational overhead.

Suppose now that we observe a stochastic process that produces $p \times 1$ vector responses or measurements $y(t)$ for

$t = 1, \ldots, n$. A signal-plus-noise model assumes that the observed process is the sum of a signal process $\{f(t) : t = 1, \ldots, n\}$ having zero mean (i.e., $\mathrm{E}[f(t)] = 0$ for $t = 1, \ldots, n$) and a zero mean noise process $\{e(t) : t = 1, \ldots, n\}$. Thus, we have

$$y(t) = f(t) + e(t), \quad t = 1, \ldots, n, \qquad (1.1)$$

where

$$\mathrm{Cov}(f(s), e(t)) = 0 \text{ for all } t, s = 1, \ldots, n, \qquad (1.2)$$

with 0 denoting a $p \times p$ matrix of all zeros, and

$$W = \{\mathrm{Cov}(e(s), e(t))\}_{s,t=1:n}$$
$$= \mathrm{diag}(W(1), \ldots, W(n)) \qquad (1.3)$$

for $p \times p$, positive definite, covariance matrices $W(1)$, $\ldots, W(n)$. Given various subsets of $\{y(1), \ldots, y(n)\}$, the goal is both point and interval prediction of the unknown random signal vectors $f(1), \ldots, f(n)$.

1.2.1   Best linear unbiased prediction

For prediction purposes we will focus on linear predictors that are unbiased and optimal in a least-squares sense. To formulate this idea suppose that z is a $r \times 1$ random vector with mean μ_z that we wish to predict using an observable $m \times 1$ random vector v that has zero-mean. We will consider linear predictors of form

$$\hat{z}(A, b) = Av + b$$

for an $r \times m$ matrix A and $r \times 1$ vector b. The *best linear unbiased predictor* (BLUP) of z is then defined to be the minimizer of

$$\mathrm{E}[z - \hat{z}(A, b)]^T [z - \hat{z}(A, b)] \qquad (1.4)$$

over all $r \times m$ matrices A and $r \times 1$ vectors b, subject to the side condition that $E[\hat{z}(A, b)] = E[z]$. The following result characterizes some of the properties of the optimal predictor.

Theorem 1.1 *If* $\mathrm{Var}(v)$ *is nonsingular, criterion (1.4) is minimized by taking* $b = \mu_z := b_{opt}$ *and*

$$A = \mathrm{Cov}(z, v)\mathrm{Var}^{-1}(v) := A_{opt}. \qquad (1.5)$$

The corresponding BLUP $\hat{z} = \hat{z}(A_{opt}, b_{opt})$ *has prediction error variance-covariance matrix*

$$E(z - \hat{z})(z - \hat{z})^T = \mathrm{Var}(z) - A_{opt}\mathrm{Var}(v)A_{opt}^T$$

$$:= V_{opt}. \qquad (1.6)$$

The BLUP is invariant with respect to location and scale changes in the sense that for any fixed $k \times r$ *matrix* B *and* $k \times 1$ *vector* c, *the BLUP of* $Bz + c$ *is* $B\hat{z} + c$ *with prediction error variance-covariance matrix* $BV_{opt}B^T$. *Moreover, if* c *is random with mean* μ_c *and is uncorrelated with* z *and* v, *the BLUP of* $Bz + c$ *is* $B\hat{z} + \mu_c$.

Proof. To establish (1.5) first observe that the unbiasedness condition entails that $b = \mu_z$. So, it suffices to restrict attention to the case where both z and v have zero means and find the matrix A that is optimal in that instance. Now note that if $\hat{z} = A_{opt}v$ and $\tilde{z} = Av$ is any other linear predictor, we have

$$E(z - \tilde{z})^T(z - \tilde{z})$$

$$= E(z - \hat{z} - [\tilde{z} - \hat{z}])^T(z - \hat{z} - [\tilde{z} - \hat{z}])$$

$$= E(z - \hat{z})^T(z - \hat{z}) - 2E(z - \hat{z})^T(\tilde{z} - \hat{z})$$

$$+ E(\tilde{z} - \hat{z})^T(\tilde{z} - \hat{z}).$$

Now let tr denote the trace functional that sums the diagonal elements of a matrix. This functional has the cyclic property that for two matrices C, D, $\mathrm{tr}[CD] = \mathrm{tr}[DC]$ assuming the products CD and DC are defined. Using this cyclic property we have

$$\mathrm{E}(z - \hat{z})^T (\tilde{z} - \hat{z})$$

$$= \mathrm{tr}\left\{ (A - A_{opt})\mathrm{E}v(z - \hat{z})^T \right\}$$

$$= \mathrm{tr}\left\{ (A - A_{opt})[\mathrm{Cov}(v, z) - \mathrm{Var}(v)A_{opt}^T] \right\} = 0,$$

which proves the first part of the theorem.

To establish location and scale invariance apply the BLUP formula to prediction of $Bz + c$ by $Av + b$ for a $k \times m$ matrix A and $k \times 1$ vector b. The optimal b is $B\mu_z + c$ and the optimal choice of A is

$$\tilde{A}_{opt} = \mathrm{Cov}(Bz + c, v)\mathrm{Var}^{-1}(v)$$

$$= B\,\mathrm{Cov}(z, v)\mathrm{Var}^{-1}(v) = BA_{opt},$$

with A_{opt} as in (1.5). The form of the variance-covariance matrix follows from the fact that $\mathrm{Var}(Bz + c - B\hat{z} - c) = \mathrm{Var}(B[z - \hat{z}])$.

Finally, if c is random we have

$$\mathrm{E}(Bz + c - (A\hat{z} + b))^T (Bz + c - (A\hat{z} + b))$$

$$= \mathrm{E}(Bz + \mu_c - A\hat{z} - b)^T (Bz + \mu_c - A\hat{z} - b)$$

$$+ 2\mathrm{E}(Bz + \mu_c - A\hat{z} - b)^T (c - \mu_c)$$

$$+ (c - \mu_c)^T (c - \mu_c).$$

The middle term in the right hand side of this expression vanishes and the result then follows from the location and scale invariance property. ∎

Theorem 1.1 provides a cornerstone for subsequent developments and we will use it, often without explicit reference, throughout the rest of the book. Some specific applications of (1.5)–(1.6) in the context of model (1.1)–(1.3) include the case where $z = f(t)$ for some fixed t

and v represents a subset of the $y(\cdot)$ process. For example, if the process has been observed up to time t, then $v = (y^T(1), \ldots, y^T(t))^T$ is a natural choice. The result concerning location/scale invariance proves useful when we subsequently consider multi-step ahead predictions and in Chapter 8 where we examine a more general model where there are nonzero means.

1.2.2 Gramm-Schmidt and innovations

In regression analysis it is often useful for interpretive and/or computational reasons to orthogonalize the predictor variables. In our setting the predictors will generally be the $y(\cdot)$ response vectors for which our particular orthogonalization scheme, detailed below, will result in the so-called *innovation process* $\varepsilon(t)$, $t = 1, \ldots, n$. To produce the innovations we simply apply the Gramm-Schmidt process (see, e.g., Davis 1975) to $y(1), \ldots, y(n)$. To accomplish this we use the fact that the $y(\cdot)$ are vector valued functions on the probability space that is implicit in the definition of model (1.1)–(1.3). In that respect it would be more complete to use notation such as $y(t, \omega)$ rather than $y(t)$ with ω being an element of the sample space Ω corresponding to the "experiment" that generated the $y(\cdot)$ process. While we have adopted a notation that suppresses the role of the probability space, it is important to remember that the orthogonality we refer to subsequently derives from thinking of the $y(t)$ as corresponding to an indexed (by t) collection of functions on Ω rather than a single functions of the "time" index t.

To initialize the Gramm-Schmidt recursion we use

$$\varepsilon(1) = y(1) \tag{1.7}$$

$$R(1) = \text{Var}(\varepsilon(1)) = \text{Var}(y(1)). \tag{1.8}$$

Then,

$$\varepsilon(t) = y(t) - \sum_{j=1}^{t-1} \text{Cov}(y(t), \varepsilon(j)) R^{-1}(j)\varepsilon(j) \qquad (1.9)$$

with

$$R(t) = \text{Var}(\varepsilon(t)) \qquad (1.10)$$

for $t = 2, \ldots, n$.

The vectors $\{\varepsilon(t) : t = 1, \ldots, n\}$ are referred to as the *innovations* or *innovation process*. One may quickly check that they are block-wise orthogonal (or merely uncorrelated in this context) in the sense that

$$\text{E}\,\varepsilon(s)\varepsilon^T(t) = 0 \text{ for } s \neq t. \qquad (1.11)$$

In addition the Gramm-Schmidt construction method insures that for each $t = 1, \ldots, n$, $\{y(1), \ldots, y(t)\}$ and $\{\varepsilon(1), \ldots, \varepsilon(t)\}$ span the same linear subspace. That is, when considered as functions on the sample space Ω, every linear combination of $y(1), \ldots, y(t)$ admits an equivalent representation as a linear combination of $\varepsilon(1), \ldots, \varepsilon(t)$ and conversely. Putting this together with (1.11) has the consequence that $\varepsilon(1), \ldots, \varepsilon(t)$ provide an orthogonal basis for the linear subspace of functions on the sample space Ω that can be written as linear combinations of the random vectors $y(1), \ldots, y(t)$.

A direct application of formulas (1.5)–(1.6) along with (1.11) reveals that the BLUP of $f(t)$ obtained from the random vectors $\{\varepsilon(1), \ldots, \varepsilon(j)\}$ has the representation

$$f(t|j) = \sum_{k=1}^{j} \text{Cov}\,(f(t), \varepsilon(k)) R^{-1}(k)\varepsilon(k) \qquad (1.12)$$

with associated prediction error variance-covariance ma-

trix

$$V(t|j) = \mathrm{E}[f(t) - f(t|j)][f(t) - f(t|j)]^T$$
$$= \mathrm{Var}\ (f(t))$$

$$-\sum_{k=1}^{j} \mathrm{Cov}\ (f(t), \varepsilon(k))R^{-1}(k)\mathrm{Cov}\ (\varepsilon(k), f(t)).$$

$$(1.13)$$

Since the random vectors $\{\varepsilon(1), \ldots, \varepsilon(j)\}$ and $\{y(1), \ldots, y(j)\}$ span the same linear space it follows that $f(t|n)$ and $V(t|n)$ agree identically with the prediction and pre-diction error variances and covariances that would have been obtained by applying (1.5)–(1.6) directly to $\{y(1), \ldots, y(j)\}$. We will eventually see that there are some computational advantages that arise in the state-space setting from using the innovations in lieu of the respons-es to formulate the prediction problem.

At this point a word or two is due on notation. While there appears to be no universal standard on this score, we will employ one of many variants on a common theme that appears in the literature. (See, e.g., Kohn and Ans-ley, 1989.) In this respect we will use a notation con-taining three parts: statistical object, time point of es-timation interest and an index that represents the part of the observed $y(\cdot)$ data to be used in constructing the object. Thus, $f(t|r)$ indicates a predictor (i.e., a statis-tic) of $f(t)$ based on the response vectors $y(1), \ldots, y(r)$. Similarly, $V(t|r)$ is used to denote its prediction error variance-covariance matrix. There is no question that no-tation can provide a formidable barrier to entry into the world of Kalman filtering. To help with this a table of the key notational objects and their definitions is pro-vided in Appendix B.

1.2.3   Gramm-Schmidt and Cholesky

Another reason for focusing on the innovations concerns
their relationship to the Cholesky matrix factorization
algorithm that is described in detail in Appendix A. To
appreciate the connection between the Gramm-Schmidt
construction of the innovations and the Cholesky method
let

$$y = (y^T(1), \ldots, y^T(n))^T$$

and write the variance-covariance matrix for y in its C-
holesky form as

$$\mathrm{Var}(y) = LRL^T \qquad (1.14)$$

with L a block lower triangular matrix having identity
matrices on the diagonal and R a block diagonal matrix.
Then, we claim that

$$R = \mathrm{diag}(R(1), \ldots, R(n)) \qquad (1.15)$$

and $L = \{L(t, j)\}_{t,j=1:n}$ with

$$L(t, j) = \begin{cases} \mathrm{Cov}\,(y(t), \varepsilon(j))R^{-1}(j) & , j = 1, \ldots, t-1, \\ I & , j = t, \\ 0 & , j > t, \end{cases}$$

$$(1.16)$$

where $R(j)$ and $\varepsilon(j)$ are defined in (1.7)–(1.10). Thus,
computation of the innovations is equivalent to computing the Cholesky factorization of $\mathrm{Var}(y)$ and conversely.

To verify our claim observe from (1.9)–(1.10) that the
vector of innovations

$$\varepsilon = (\varepsilon^T(1), \ldots, \varepsilon^T(n))^T$$

can be expressed as

$$\varepsilon = y + A\varepsilon$$

with $A = I - L$ for L defined in (1.16). Thus, $L\varepsilon = y$ and, since the $\varepsilon(j)$ are block-wise uncorrelated with $\mathrm{Var}(\varepsilon) = R$, the identity $\mathrm{Var}(y) = LRL^T$ is established. The uniqueness of the Cholesky decomposition completes the proof.

The connection between the Cholesky factorization and the Gramm-Schmidt orthogonalization method has important computational implications. To see this let

$$f = (f^T(1), \ldots, f^T(n))^T$$

$$e = (e^T(1), \ldots, e^T(n))^T,$$

and observe that

$$\widehat{f} = \mathrm{Cov}(f, y)\mathrm{Var}(y)^{-1}y$$

$$= \mathrm{Cov}(y - e, y)\mathrm{Var}(y)^{-1}y$$

$$= y - W\mathrm{Var}(y)^{-1}y$$

$$= y - W(L^T)^{-1}R^{-1}L^{-1}y$$

$$= y - W(L^T)^{-1}R^{-1}\varepsilon. \tag{1.17}$$

In establishing this we employed the identity $f = y - e$ and the fact that $\mathrm{Cov}(e, y) = W$ since f and e are uncorrelated. A similar calculation produces

$$V = \mathrm{Var}(f - \widehat{f})$$

$$= \mathrm{Var}(f) - \mathrm{Cov}(f, y)\mathrm{Var}(y)^{-1}\mathrm{Cov}(y, f)$$

$$= W - W(L^T)^{-1}R^{-1}L^{-1}W \tag{1.18}$$

with $V(t|n)$, $t = 1, \ldots, n$, being the diagonal blocks of this matrix. Consequently, an efficient implementation of the Cholesky/Gramm-Schmidt recursion is one path that can be taken toward developing computational methods for prediction in model (1.1).

1.3   State-space models

Let us now specialize the previous development to the case where the measurement vectors $y(t)$ in (1.1) follow a *state-space model*. By this we mean that the signal vectors may be represented as

$$f(t) = H(t)x(t),$$

for $q \times 1$ unobserved *state* vectors $x(t)$, $t = 1, \ldots, n$, and known matrices $H(t)$, $t = 1, \ldots, n$. The state vectors are assumed to propagate via the *state equations*

$$x(t+1) = F(t)x(t) + u(t)$$

with $F(t)$, $t = 0, \ldots, n-1$, known $q \times q$ matrices and $u(0), \ldots, u(n-1)$ unobservable, zero mean random perturbations. There is also an initial state vector $x(0)$ that initializes the process.

By combining all the above specifications we can formally set out the state-space model as:

$$y(t) = H(t)x(t) + e(t), \quad t = 1, \ldots, n, \qquad (1.19)$$

and

$$x(t+1) = F(t)x(t) + u(t), \quad t = 0, \ldots, n-1, \qquad (1.20)$$

where the $e(t)$, $u(t-1)$, $t = 1, \ldots, n$, and $x(0)$ all have zero means and the model's covariance structure is determined by the conditions

$$\mathrm{Cov}(e(t), e(s)) = 0, \; s \neq t, \qquad (1.21)$$

$$\mathrm{Var}(e(t)) = W(t), \; t = 1, \ldots, n, \qquad (1.22)$$

for positive-definite matrices $W(t)$, $t = 1, \ldots, n$,

$$\text{Cov}(u(s), u(t)) \;=\; 0, \; s \neq t, \tag{1.23}$$

$$\text{Var}(u(t)) \;=\; Q(t), \; t = 0, \ldots, n - 1, \tag{1.24}$$

for positive semi-definite matrices $Q(t)$, $t = 0, \ldots, n-1$,

$$\text{Var}(x(0)) \;=\; S(0|0), \tag{1.25}$$

and, for $t = 1, \ldots, n$ and $s = 0, \ldots, n - 1$,

$$\text{Cov}(e(t), u(s)) \;=\; 0, \tag{1.26}$$

$$\text{Cov}(e(t), x(0)) \;=\; 0, \tag{1.27}$$

$$\text{Cov}(u(s), x(0)) \;=\; 0. \tag{1.28}$$

The presence of the initial state vector $x(0)$ is actually somewhat problematic. While there are situations where $S(0|0)$ is known, there are many instances where this is not the case which causes practical difficulties in initializing the model. The problem can be resolved in ways that are discussed in Chapter 6. For now we will ignore this aspect of the state-space formulation and proceed as if $S(0|0)$ in (1.25) is known.

In the state-space model least-squares estimation of the signal $f(t)$ becomes equivalent to estimation of the state vector $x(t)$. In this regard the BLUP of $x(t)$ based on $\varepsilon(1), \ldots, \varepsilon(j)$ is

$$x(t|j) \;=\; \sum_{k=1}^{j} \text{Cov}\,(x(t), \varepsilon(k)) R^{-1}(k) \varepsilon(k) \tag{1.29}$$

with associated prediction error variance-covariance ma-

trix

$$S(t|j) = \mathrm{E}[x(t) - x(t|j)][x(t) - x(t|j)]^T$$

$$= \mathrm{Var}\,(x(t))$$

$$- \sum_{k=1}^{j} \mathrm{Cov}\,(x(t), \varepsilon(k))R^{-1}(k)\mathrm{Cov}\,(\varepsilon(k), x(t)).$$

$$(1.30)$$

Then, either directly or by using the linear transformation invariance of BLUPs, we see that

$$f(t|j) = H(t)x(t|j)$$

and

$$V(t|j) = H(t)S(t|j)H^T(t)$$

are the BLUP and prediction error variance-covariance matrix for estimation of $f(t)$ from $\varepsilon(1), \ldots, \varepsilon(j)$.

Consider now the problem of predicting $x(t + k)$ (or $f(t + k)$) given data up to time index t.

Proposition 1.1 *For* $k \geq 1$, *the BLUP of* $x(t + k)$ *from* $\varepsilon(1), \ldots, \varepsilon(t)$ *is* $x(t + k|t) = F(t + k - 1) \cdots F(t)x(t|t)$.

Proof. Let us begin with $k = 1$ and write $x(t + 1) = F(t)x(t) + u(t)$ with $u(t)$ being uncorrelated with $x(t)$ and $\varepsilon(1), \ldots, \varepsilon(t)$. The invariance of the BLUP for $x(t)$ under a scale change and random (zero mean) location shift detailed in Theorem 1.1 can then be applied to see that the BLUP of $x(t + 1)$ is $F(t)x(t|t)$. One may now proceed by induction to verify the general conclusion of the theorem.   ∎

Proposition 1.1 has the consequence that a BLUP $x(t|t)$ at any given time index t can be easily modified for use in the prediction of a state vector at some future index values. A similar conclusion applies to estimation of the signal since

$$f(t + k \mid t) = H(t + k)x(t + k \mid t)$$

$$= H(t + k)F(t + k - 1) \cdots F(t)x(t \mid t).$$

There are many time series models that admit state-space representations. However, for our purposes it will suffice to concentrate primarily on the following two examples.

Example: AR and MA Processes. A time series $y(\cdot)$ is said to be a mth order moving average (MA) process if there are known $p \times p$ matrices $A(1), \ldots, A(m)$ such that

$$y(t) = \sum_{j=1}^{m} A(j)e(t - j) + e(t), \quad t = 1, \ldots, n, \quad (1.31)$$

for block-wise uncorrelated, zero mean, random p-vectors $e(1), \ldots, e(n)$ with common $p \times p$ covariance matrix

$$W_0 = W(1) = \cdots = W(n)$$

and initializing vector

$$x(0) = \begin{bmatrix} e(-1) \\ e(-2) \\ . \\ \vdots \\ . \\ e(-m) \end{bmatrix} = 0.$$

To write (1.31) in state-space form define $mp \times 1$ state vectors by

$$x(t) = \begin{bmatrix} e(t - 1) \\ \vdots \\ e(t - m) \end{bmatrix}.$$

Then, for $t = 0, \ldots, n - 1,$

$$x(t + 1) = F_{MA} x(t) + u(t)$$

with F_{MA} the $mp \times mp$ matrix

$$F_{MA} := \begin{bmatrix} 0 & 0 & \cdots & 0 & 0 \\ I & 0 & \cdots & 0 & 0 \\ 0 & I & \cdots & 0 & 0 \\ & \cdot & & \cdot & \\ & \cdot & & \cdot & \\ & \cdot & & \cdot & \\ 0 & 0 & \cdots & I & 0 \end{bmatrix} \qquad (1.32)$$

and $u(t) = T(m)e(t)$ for $T(m)$ the $mp \times p$ matrix

$$T(m) := \begin{bmatrix} I \\ 0 \\ \cdot \\ \cdot \\ \cdot \\ 0 \end{bmatrix}. \qquad (1.33)$$

Thus, the state-space representation holds with

$$H(t) = [A(1), A(2), \ldots, A(m)] := H_{MA},$$

for $t = 1, \ldots, n$ and

$$F(t) = F_{MA},$$

$$Q(t) = T(m)W_0 T^T(m)$$

$$= \begin{bmatrix} W_0 & 0 & \cdots & 0 \\ 0 & 0 & \cdots & 0 \\ \cdot & \cdot & \cdot & \cdot \\ \cdot & \cdot & & \cdot \\ \cdot & \cdot & & \cdot \\ 0 & 0 & \cdots & 0 \end{bmatrix}$$

$$:= Q_{MA}$$

for $t = 0, \ldots, n - 1$.

An rth order autoregressive (AR) process has the form

$$y(t) = \sum_{j=1}^{r} B(j)y(t - j) + e(t)$$

with $B(1), \ldots, B(r)$ known coefficient matrices, the $e(\cdot)$ zero mean, block wise uncorrelated random vectors and

$$y(-1) = \cdots = y(-r) = 0.$$

In this case we can take

$$x(t) = \begin{bmatrix} y(t-1) \\ \vdots \\ y(t-r) \end{bmatrix}$$

giving $x(0) = 0$ and

$$x(t+1) = F_{AR}x(t) + T(r)e(t)$$

for

$$F_{AR} := \begin{bmatrix} B(1) & B(2) & \cdots & B(r-1) & B(r) \\ I & 0 & \cdots & 0 & 0 \\ 0 & I & \cdots & 0 & 0 \\ \vdots & & & & \\ 0 & 0 & \cdots & I & 0 \end{bmatrix},$$

$$H(t) = [B(1), \ldots, B(r)] := H_{AR},$$

$T(r)$ defined as in (1.33) for the MA case and

$$Q(t) = T(r)W_0 T^T(r) := Q_{AR}.$$

By combining the MA and AR models we obtain an ARMA or autoregressive moving average process with

$$y(t) = \sum_{j=1}^{r} B(j)y(t-j) + \sum_{j=1}^{m} A(j)e(t-j) + e(t).$$

This has a state-space representation with state vector

$$x(t) = \begin{bmatrix} y(t-1) \\ \vdots \\ y(t-m) \\ e(t-1) \\ \vdots \\ e(t-m) \end{bmatrix},$$

satisfying $x(0) = 0$ and

$$F_{ARMA} := [F_{AR}, F_{MA}]$$

$$u(t) = \begin{bmatrix} T(r) \\ T(m) \end{bmatrix} e(t),$$

$$Q(t) = \begin{bmatrix} T(r) \\ T(m) \end{bmatrix} W_0 \begin{bmatrix} T^T(r) & T^T(m) \end{bmatrix}$$

$$:= Q_{ARMA},$$

$$H(t) = [H_{AR}, H_{MA}] := H_{ARMA}.$$

A key feature of the MA, AR and ARMA models is that their associated $F(t)$, $Q(t)$, $H(t)$ and $W(t)$ matrices are constant as a function of t. Thus, when we subsequently treat examples where $H(t)$, $F(t)$, $Q(t)$ and $W(t)$ are time independent, this will have direct applications to AR, MA and ARMA model settings.

Example: Brownian Motion with White Noise. One way to obtain realizations of a discrete time signal-plus-noise process is to sample from a continuous time process such as

$$z(\tau) = B(\tau) + v(\tau), \quad \tau \in [0, 1], \tag{1.34}$$

for $v(\cdot)$ and $B(\cdot)$ zero mean processes that are uncorrelated with one another. In this respect we will be particularly interested in the case where $v(\cdot)$ is a zero mean process with covariances determined by

$$\text{Cov}(v(\tau), v(\nu)) = \begin{cases} W_0, & \nu = \tau, \\ 0, & \nu \neq \tau \end{cases} \tag{1.35}$$

and $B(\cdot)$ is a zero mean process with covariance kernel

$$\text{Cov}(B(\tau), B(\nu)) = \min(\tau, \nu) \tag{1.36}$$

When $B(\cdot)$ is a normal process it is referred to as *Brownian motion*. Zero mean processes with covariance functions of the form (1.35) are frequently referred to as *white noise*.

Suppose that we sample from (1.34) at points $0 \leq \tau_1 < \cdots < \tau_n \leq 1$. Then, we see that the resulting measurements are generated as in (1.1) upon making the identifications $y(t) = z(\tau_t)$, $f(t) = B(\tau_t)$ and $e(t) = v(\tau_t)$. To obtain the state-space formulation for the responses we can take $x(t) = B(\tau_t)$ thereby giving $H(t) \equiv 1$. Since $B(0) \equiv 0$, the initial state value is $x(0) \equiv 0$.

For the state equation we use

$$x(t) = x(t-1) + u(t-1)$$

with $u(t-1) = B(\tau_t) - B(\tau_{t-1})$, since $\text{Cov}(u(i), u(j)) = 0$ for $i \neq j$. This means that $F(t) \equiv 1$, $Q(0) = \tau_1$ and, for $t = 1, \ldots, n-1$,

$$Q(t) = \tau_{t+1} - \tau_t.$$

An illustration of the type of data being considered in this example is shown in Figure 1.1. Here we see the

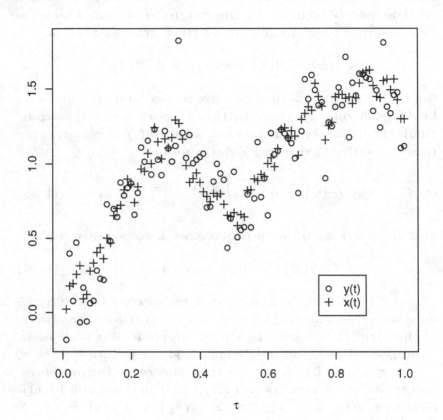

FIGURE 1.1

Sampling from Brownian motion

actual values of the unobserved Brownian motion sample path at the τ_j (indicated by "+" symbols) recorded without the white noise perturbations. The actual sampled responses/measurements which include the white noise disturbances are represented by circles. The samples were taken at $n = 100$ equally spaced points $\tau_t = t/100$, $t = 1, \ldots, 100$ with the added noise variables having variances $W_0 = .025$. The goal in this particular case would be to "filter out" the white noise from the responses in an attempt to recover the actual values from the Brownian motion sample path that occurred at the sampling point.

1.4 What lies ahead

In the chapters that follow we study computational methods for obtaining the quantities $f(t|j)$, $V(t|j)$, $x(t|j)$ and $S(t|j)$ in (1.12)–(1.13) and (1.29)–(1.30) in the state-space setting. As a result of our expansions for $f(t|j)$ and $x(t|j)$ using the orthogonal innovation basis vectors, this is tantamount to efficiently computing the covariance matrices $\mathrm{Cov}(f(t), \varepsilon(k))$, $\mathrm{Cov}(x(t), \varepsilon(k))$, $R(k)$, $k = 1$, $\ldots, j$ as well as the innovations $\varepsilon(1), \ldots, \varepsilon(k)$. The common component in all these factors is the innovation vectors whose computation is linked directly to the Cholesky factorization of $\mathrm{Var}(y)$. Consequently, the Cholesky decomposition is the unifying theme for all that follows and is the perspective we will adopt for viewing developments throughout the text.

In the next chapter we lay the groundwork for Chapters 3–5 by establishing the essential recurrence structure for the covariances $\mathrm{Cov}(x(t), \varepsilon(k))$ that arise in the state vector BLUPs (1.29) and their associated prediction error variance-covariance matrices (1.30). Then, in Chapter 3, we show how this structure can be exploited to obtain a computationally efficient, modified Cholesky factorization of $\mathrm{Var}(y)$ as well as $\mathrm{Var}^{-1}(y)$.

Chapters 4-5 use the results in Chapters 2–3 to derive the forward and backward (i.e., the smoothing step) Kalman filter recursions. In the case of signal estimation, we will see that these are basically straightforward consequences of the efficient Cholesky factorization for $\mathrm{Var}(y)$ that becomes possible under a state-space formulation.

Chapter 6 deals with the problem of specifying the distribution (or value) for the initial state vector $x(0)$. One way to circumvent this problem is to employ a diffuse specification which leads to the diffuse Kalman filter that we study in some detail.

In the case where $y(\cdot)$ and $x(\cdot)$ are normal processes, the observed responses can be used to obtain a sample likelihood that can be useful for inference about unknown model parameters. We show how the Kalman filter can be used to efficiently evaluate the likelihood for standard as well as diffuse specifications for the distribution of $x(0)$ in Chapter 7. Finally, to complete our treatment of the Kalman filter, a more general state–space model is introduced in Chapter 8 and we describe how the results from the previous chapters extend to this case.

2

The Fundamental Covariance
Structure

2.1 Introduction

In this chapter we will lay the foundation for Chapters
3–5. In that regard, our goal is to obtain a complete characterization of the covariance relationship between the
innovations and the state vectors. Providing a detailed
exposition of this structure is the focus of Section 2.3.
First, however, we develop several tools that are needed to prove Lemma 2.4 in Section 2.3 as well as other
results in subsequent chapters.

2.2 Some tools of the trade

To begin let us recall some of the basic formulations from
Chapter 1. First, the assumption is that we have responses given by

$$y(t) = H(t)x(t) + e(t) \tag{2.1}$$

with the state vectors developing from the recursion

$$x(t + 1) = F(t)x(t) + u(t). \tag{2.2}$$

Here $e(1), \ldots, e(n), u(0), \ldots, u(n-1), x(0)$ are all zero mean, uncorrelated random vectors having $\text{Var}(e(j)) = W(j), j = 1, \ldots, n$, $\text{Var}(u(j)) = Q(j), j = 0, \ldots, n-1$ and $\text{Var}(x(0)) = S(0|0)$. The responses are then transformed into the innovations via the Gramm-Schmidt recursion:

$$\varepsilon(1) = y(1),$$

$$R(1) = \text{Var}(\varepsilon(1)) = \text{Var}(y(1))$$

and

$$\varepsilon(t) = y(t) - \sum_{j=1}^{t-1} \text{Cov}(y(t), \varepsilon(j)) R^{-1}(j)\varepsilon(j) \qquad (2.3)$$

with

$$R(t) = \text{Var}(\varepsilon(t)) \qquad (2.4)$$

for $t = 2, \ldots, n$.

From the innovations we obtain the BLUP of $x(t)$ from $y(1), \ldots, y(j)$ as

$$x(t|j) = \sum_{i=1}^{j} \text{Cov}(x(t), \varepsilon(i)) R^{-1}(i)\varepsilon(i) \qquad (2.5)$$

with associated prediction error variance-covariance matrix

$$\begin{aligned}
S(t|j) &= \text{Var}[x(t) - x(t|j)] \\
&= \text{Var}(x(t)) \\
&\quad - \sum_{i=1}^{j} \text{Cov}(x(t), \varepsilon(i)) R^{-1}(i)\text{Cov}(\varepsilon(i), x(t)) \\
&= \text{Var}(x(t)) - \text{Var}(x(t|j)). \qquad (2.6)
\end{aligned}$$

Note that

$$\text{Cov}(x(t), x(t|j))$$

$$= \sum_{i=1}^{j} \text{Cov}(x(t), \varepsilon(i)) R^{-1}(i) \text{Cov}(\varepsilon(i), x(t))$$

$$= \text{Var}(x(t|j)).$$

Thus, $x(t|j)$ has the orthogonality property

$$\text{Cov}(x(t) - x(t|j), x(t|j)) = 0 \qquad (2.7)$$

or, equivalently,

$$S(t|j) = \text{Cov}(x(t) - x(t|j), x(t)). \qquad (2.8)$$

We will now begin our study of the relationships between the innovations, least squares predictors and state vectors. For this purpose it will be worthwhile to first set out a few important facts that will arise again and again in our study of state-space models. Specifically, we can see that the following properties are immediate consequences of the state-space model assumptions and the definition of the innovations:

(F1) $\varepsilon(t)$ is uncorrelated with $\varepsilon(s)$, $s \neq t$ and $e(s)$ for $s > t$.

(F2) $e(t)$ is uncorrelated with $x(s)$ for all s.

(F3) $u(t)$ is uncorrelated with $x(s)$, $y(s)$, $\varepsilon(s)$ for $s \leq t$.

As an illustration of the use of (F1), note that by combining (F1), (2.1)–(2.2) and (2.3) we obtain

$$\varepsilon(t) \;=\; y(t) - \sum_{j=1}^{t-1} \mathrm{Cov}(y(t),\,\varepsilon(j)) R^{-1}(j)\varepsilon(j)$$

$$= H(t)x(t) + e(t)$$

$$- \sum_{j=1}^{t-1} \mathrm{Cov}(H(t)x(t) + e(t),\,\varepsilon(j)) R^{-1}(j)\varepsilon(j)$$

$$= H(t)[x(t) - \sum_{j=1}^{t-1} \mathrm{Cov}(x(t),\,\varepsilon(j)) R^{-1}(j)\varepsilon(j)]$$

$$+ e(t)$$

$$= H(t)[x(t) - x(t|t-1)] + e(t). \qquad (2.9)$$

Many of the developments that follow depend on our being able to efficiently compute the quantities $S(t|t-1)$, $S(t|t)$ in (2.6) and $R(t)$ in (2.4). We conclude this section by establishing recursive formulae for each of these three quantities.

Lemma 2.1 *Let*

$$S(1|0) := \mathrm{Var}(x(1)). \qquad (2.10)$$

Then,

$$S(t|t) = S(t|t-1)$$

$$- S(t|t-1)H^{T}(t) R^{-1}(t) H(t) S(t|t-1)$$

$$(2.11)$$

for $t = 1, \ldots, n$.

Proof. From (2.6)

$$S(t|t) = S(t|t-1)$$

$$-\text{Cov}(x(t), \varepsilon(t))R^{-1}(t)\text{Cov}(\varepsilon(t), x(t)).$$

We can now use (2.8)–(2.9) along with (F2) to complete the proof. ∎

Lemma 2.2 *Define $S(1|0)$ as in (2.10). Then, for $t = 1$, ..., n,*

$$S(t|t-1) = F(t-1)S(t-1|t-1)F^T(t-1)$$

$$+Q(t-1). \qquad (2.12)$$

Proof. Proposition 1.1 has the consequence that for $t = 2, \ldots, n$

$$x(t) - x(t|t-1)$$

$$= F(t-1)x(t-1) + u(t-1)$$

$$-F(t-1)x(t-1|t-1)$$

$$= F(t-1)[x(t-1) - x(t-1|t-1)]$$

$$+u(t-1). \qquad (2.13)$$

The result now follows from the definition of $S(t-1|t-1)$ in (2.6) and (F3). ∎

Lemma 2.3 *Let $S(1|0)$ be defined as in (2.10). Then,*

$$R(t) = H(t)S(t|t-1)H^T(t) + W(t) \qquad (2.14)$$

for $t = 1, \ldots, n$.

Proof. First note that

$$R(1) \ = \ \mathrm{Var}(\varepsilon(1)) = \mathrm{Var}(y(1)) = \mathrm{Var}[H(1)x(1) + e(1)]$$

$$= \ H(1)\mathrm{Var}(x(1))H^T(1) + W(1)$$

due to (F2). The result for $t = 2, \ldots, n$ is immediate from (2.9) and (F2).   ∎

As a consequence of Lemmas 2.1-2.3 we can recursively update $S(t|t-1)$, $S(t|t)$ and $R(t)$ via the following algorithm.

Algorithm 2.1 This algorithm computes $S(t|t-1)$, $R(t)$ and $S(t|t)$ for $t = 1, \ldots n$.

> /*Initialization*/
> $S(1|0) = F(0)S(0|0)F^T(0) + Q(0)$
> $R(1) = H(1)S(1|0)H^T(1) + W(1)$
> $S(1|1) = S(1|0) - S(1|0)H^T(1)R^{-1}(1)H(1)S(1|0)$
> **for** $t = 2$ **to** n
> $S(t|t-1) = F(t-1)S(t-1|t-1)F^T(t-1)$
> $+Q(t-1)$
> $R(t) = H(t)S(t|t-1)H^T(t) + W(t)$
> $S(t|t) = S(t|t-1)$
> $-S(t|t-1)H^T(t)R^{-1}(t)H(t)S(t|t-1)$
> **end for**

Note from Algorithm 2.1 that for each t the amount of work involved in evaluating $S(t|t-1)$, $S(t|t)$ and $R(t)$ depends only on the dimensions of the matrices and not on n. Thus, assuming n to be the dominant computational factor, the above recursion returns the entire collection of matrices $S(1|0)$, $R(1)$, $S(1|1)$, $\ldots$, $S(n|n-1)$, $R(n)$, $S(n|n)$ in a total of order n floating point operations or *flops*.

2.3 State and innovation covariances

The stage has now been set to accomplish the goals of this chapter. Specifically, we are ready to derive a recursive characterization for the quantities $\text{Cov}(x(t), \varepsilon(j))$ that appear, for example, in the BLUP for $x(t)$ in (2.5) and its associated prediction error covariance matrix in (2.6). We state the result formally in Lemma 2.4 below.

Lemma 2.4 *Let*

$$S(1|0) := \text{Var}(x(1)) = F(0)S(0|0)F^T(0) + Q(0).$$

Then, for $t = 1, \ldots, n$,

$$\text{Cov}(x(t), \varepsilon(t)) = S(t|t-1)H^T(t) \qquad (2.15)$$

and, for $j \leq t - 1$,

$$\text{Cov}(x(t), \varepsilon(j)) = F(t-1) \cdots F(j)S(j|j-1)H^T(j). \qquad (2.16)$$

Let

$$M(t) = F(t) - F(t)S(t|t-1)H^T(t)R^{-1}(t)H(t). \qquad (2.17)$$

Then, for $t = n-1, \ldots, 1$ and $j \geq t+1$,

$$\text{Cov}(x(t), \varepsilon(j))$$

$$= S(t|t-1)M^T(t)M^T(t+1) \cdots M^T(j-1)H^T(j). \qquad (2.18)$$

This lemma provides a key for developing recursive computational schemes for evaluating the $nq \times np$ covariance matrix

$$\Sigma_{X\varepsilon} = \{\sigma_{X\varepsilon}(t, j)\}_{t,j=1:n}$$
$$= \{\mathrm{Cov}(x(t), \varepsilon(j))\}_{t,j=1:n} \qquad (2.19)$$

as well as the BLUPs of the signal and state vectors. For the moment we will concentrate on efficiently computing $\Sigma_{X\varepsilon}$. By efficient we mean a small number of flops and in this respect the very best we could generally hope for would be to evaluate all of $\Sigma_{X\varepsilon}$ in order n^2 operations in cases where p and q are "small" relative to n. Clearly, such a recursion is only possible if the calculation of each element of $\Sigma_{X\varepsilon}$ can be accomplished in an amount of computational effort that is independent of n. We will see that this is, in fact, the case for the state-space setting.

To begin exploring the implications of Lemma 2.4 let us consider how to efficiently calculate the diagonal and below diagonal entries of $\Sigma_{X\varepsilon}$. Using Lemma 2.4 we see that the first two columns of $\Sigma_{X\varepsilon}$ can be written as

$$S(1|0)H^T(1)$$
$$F(1)S(1|0)H^T(1)$$
$$F(2)F(1)S(1|0)H^T(1)$$

$$\vdots$$

$$F(n-2)\cdots F(1)S(1|0)H^T(1)$$
$$F(n-1)\cdots F(1)S(1|0)H^T(1)$$

and

$$S(1|0)M^T(1)H^T(2)$$
$$S(2|1)H^T(2)$$
$$F(2)S(2|1)H^T(2)$$

$$\vdots$$

$$F(n-2)\cdots F(2)S(2|1)H^T(2)$$
$$F(n-1)\cdots F(2)S(2|1)H^T(2)$$

By next considering the third column, then the fourth, etc., we can detect several patterns in the elements of $\Sigma_{X\varepsilon}$. In particular, we see that the diagonal and below diagonal blocks for the first two columns of $\Sigma_{X\varepsilon}$ have a progressive nature that appears like

$$S(1|0)H^T(1)$$

$$\times F(1) \downarrow$$

$$F(1)S(1|0)H^T(1)$$

$$\times F(2) \downarrow$$

$$F(2)F(1)S(1|0)H^T(1)$$

$$\times F(3) \downarrow$$

$$\vdots$$

$$\times F(n-2) \downarrow$$

$$F(n-2)\cdots F(1)S(1|0)H^T(1)$$

$$\times F(n-1) \downarrow$$

$$F(n-1)\cdots F(1)S(1|0)H^T(1)$$

and

$$S(2|1)H^T(2)$$

$$\times F(2) \Bigg\downarrow$$

$$F(2)S(2|1)H^T(2)$$

$$\times F(3) \Bigg\downarrow$$

$$\vdots$$

$$\times F(n-2) \Bigg\downarrow$$

$$F(n-2)\cdots F(2)S(2|1)H^T(2)$$

$$\times F(n-1) \Bigg\downarrow$$

$$F(n-1)\cdots F(2)S(2|1)H^T(2)$$

By extrapolating from what we have observed in these special cases we can determine that the diagonal and below diagonal blocks of $\Sigma_{X\varepsilon}$ can be computed on a row-by-row basis by simply "updating" entries from previous rows through pre-multiplication by an appropriate $F(\cdot)$ matrix. That is, we can use the elements of the tth block row of $\Sigma_{X\varepsilon}$ to evaluate all the (strictly) below diagonal blocks of the $(t+1)$st row block through pre-multiplication by $F(t)$. The $(t+1)$st or diagonal column block is then obtained from Lemma 2.4 as $\mathrm{Cov}(x(t+$

1), $\varepsilon(t+1)) = S(t+1|t)H^T(t+1)$. The following provides an algorithmic implementation of these ideas.

Algorithm 2.2 This algorithm computes the diagonal and below diagonal blocks of $\Sigma_{X\varepsilon}$.

$\sigma_{X\varepsilon}(1,1) = S(1|0)H^T(1)$
for $t = 2$ **to** n
    for $j = 1$ **to** $t-1$
        $\sigma_{X\varepsilon}(t,j) = F(t-1)\sigma_{X\varepsilon}(t-1,j)$
    end for
    $\sigma_{X\varepsilon}(t,t) = S(t|t-1)H^T(t)$
end for

Algorithm 2.2 is an example of a *forward recursion* in the sense that it works its way from the upper left hand corner of $\Sigma_{X\varepsilon}$ down to the lower right hand block of the matrix. We should also note that there is nothing special about evaluation of $\Sigma_{X\varepsilon}$ on a row-by-row basis. The entire first column block could have been evaluated, then the diagonal and below diagonal blocks of the second column block, etc. Our row-by-row approach is motivated by developments in later chapters where the elements of $\Sigma_{X\varepsilon}$ must be evaluated in a manner that is dictated by the temporal sequence of the data's collection.

Upon examining Algorithm 2.2 we see that the tth step of the recursion involves multiplications of only $q \times q$ and $q \times p$ matrices. Consequently, the overall computing effort that is needed to evaluate the diagonal and below diagonal blocks of $\Sigma_{X\varepsilon}$ is $O(n^2)$ flops provided that p, q are small relative to n.

Now let us consider how one might compute the above diagonal entries for $\Sigma_{X\varepsilon}$. Similar to what we did for the below diagonal entries let us start by using Lemma 2.4 to write out the last two columns for $\Sigma_{X\varepsilon}$. Column $(n-1)$

is seen to have the form

$$S(1|0)M^T(1) \cdots M^T(n-2)H^T(n-1)$$
$$S(2|1)M^T(2) \cdots M^T(n-2)H^T(n-1)$$

$$\vdots$$

$$S(n-3|n-4)M^T(n-3)M^T(n-2)H^T(n-1)$$
$$S(n-2|n-3)M^T(n-2)H^T(n-1)$$
$$S(n-1|n-2)H^T(n-1)$$
$$F(n-1)S(n-1|n-2)H^T(n-1)$$

while the blocks in column n are

$$S(1|0)M^T(1) \cdots M^T(n-1)H^T(n)$$
$$S(2|1)M^T(2) \cdots M^T(n-1)H^T(n)$$

$$\vdots$$

$$S(n-3|n-4)M^T(n-3) \cdots M^T(n-1)H^T(n)$$
$$S(n-2|n-3)M^T(n-2)M^T(n-1)H^T(n)$$
$$S(n-1|n-2)M^T(n-1)H^T(n)$$
$$S(n|n-1)H^T(n)$$

The pattern that appears here is similar to what we saw in working out the forward recursion for the diagonal and below diagonal elements of $\Sigma_{X\varepsilon}$ in the sense that each row can be updated (or "downdated") as we move (backward) to the row above through the use of a common pre-multiplier. If we set $A(t, t) = H^T(t)$, then we need to update and retain the matrices

$$A(t, j) = M^T(t) \cdots M^T(j-1)H^T(j), \ j = t+1, \ldots, n,$$

that appear in the above diagonal column blocks for the tth row. These matrices can then be used to compute the new row blocks on each backward step with the updating accomplished via the relation $A(t-1, j) = M^T(t-1)A(t, j)$. The resulting matrices are then pre-multiplied by $S(t-1|t-2)$ to obtain the blocks in the $(t-1)$th row of $\Sigma_{X\varepsilon}$. In particular, for our two special cases this updating process can be depicted as

$$S(1|0)A(1, n-1)$$

$$M^T(1)A(2, n-1)$$

$$S(2|1)A(2, n-1)$$

$$M^T(2)A(3, n-1)$$

$$\vdots$$

$$M^T(n-4)A(n-3, n-1)$$

$$S(n-3|n-4)A(n-3, n-1)$$

$$M^T(n-3)A(n-2, n-1)$$

$$S(n-2|n-3)A(n-2, n-1)$$

$$M^T(n-2)A(n-1, n-1)$$

$$S(n-1|n-2)A(n-1, n-1)$$

$$F(n-1)S(n-1|n-2)H^T(n-1)$$

for the blocks in column $n - 1$ and

$$S(1|0)A(1, n)$$

$$M^T(1)A(2,n)$$

$$S(2|1)A(2, n)$$

$$M^T(2)A(3,n)$$

$$\vdots$$

$$M^T(n-4)A(n-3,n)$$

$$S(n - 3|n - 4)A(n - 3, n)$$

$$M^T(n-3)A(n-2,n)$$

$$S(n - 2|n - 3)A(n - 2, n)$$

$$M^T(n-2)A(n-1,n)$$

$$S(n - 1|n - 2)A(n - 1, n)$$

$$M^T(n-1)A(n,n)$$

$$S(n|n - 1)A(n, n)$$

for the blocks in column n. An algorithmic description of the above considerations is as follows.

Algorithm 2.3 This algorithm computes the above diagonal blocks of $\Sigma_{X\varepsilon}$.

$$A(n) = M^T(n-1)H^T(n)$$
$$\sigma_{X\varepsilon}(n-1, n) = S(n-1|n-2)A(n)$$
for $t = n - 2$ **to** 1
$$\quad A(t+1) = M^T(t)H^T(t+1)$$
$$\quad \sigma_{X\varepsilon}(t, t+1) = S(t|t-1)A(t+1)$$
$\quad$**for** $k = (t+2)$ **to** n
$$\quad\quad A(k) := M^T(t)A(k)$$
$$\quad\quad \sigma_{X\varepsilon}(t, k) = S(t|t-1)A(k)$$
$\quad$**end for**
end for

In contrast to Algorithm 2.2, Algorithm 2.3 is an example of a *backward recursion*. In this case computations proceed in "reverse" starting with a block matrix in the lower right hand corner of $\Sigma_{X\varepsilon}$ (i.e., the $(n-1)$st row and nth column block) and ending with an upper right hand block (i.e., the first row and second column block) of the matrix.

Arguing as we did for Algorithm 2.2, we see that $O(n^2)$ flops will also be needed to evaluate the upper diagonal blocks of $\Sigma_{X\varepsilon}$. Accordingly, the entire matrix can be obtained in order n^2 operations.

Proof of Lemma 2.4. First, note that (2.15) follows immediately from (2.8)–(2.9) and (F2). For (2.16) with $j \leq t-1$ we then use (2.2) to see that

$$x(t) = F(t-1) \cdots F(j)x(j) + Z(t)$$

with $Z(t)$ depending only on $u(t-1), \ldots, u(j)$. In light of (F3), Z must be uncorrelated with $\varepsilon(j)$ and the result then follows from (2.15).

The backward recursion (2.18) is somewhat more difficult to establish. First use (2.9) and (F2) followed by (2.13), (F3) and, finally, (2.8) and (2.11) to see that

$$\text{Cov } (x(t), \varepsilon(t+1))$$

$$= \text{Cov } (x(t), x(t+1) - x(t+1|t))H^T(t+1)$$

$$= \text{Cov } (x(t), x(t) - x(t|t))F^T(t)H^T(t+1)$$

$$= S(t|t)F^T(t)H^T(t+1)$$

$$= S(t|t-1)M^T(t)H^T(t+1)$$

which proves (2.18) for $j = t + 1$.

By exactly the same process we used for $j = t + 1$ we find that $\text{Cov}(x(t), \varepsilon(t+2))$ has the form

$$\text{Cov}(x(t), x(t+1) - x(t+1|t+1))F^T(t+1)H^T(t+2).$$

Now, by definition, (F3) and (2.15) we can express $x(t+1) - x(t+1|t+1)$ as

$$x(t+1) - \sum_{j=1}^{t} \text{Cov}(x(t+1), \varepsilon(j))R^{-1}(j)\varepsilon(j)$$

$$-\text{Cov}(x(t+1), \varepsilon(t+1))R^{-1}(t+1)\varepsilon(t+1)$$

$$= F(t)[x(t) - x(t|t)]$$

$$-S(t+1|t)H^T(t+1)R^{-1}(t+1)\varepsilon(t+1) + u(t).$$

Thus, using (F3), (2.8), the definition of $M(t)$ and our previous result for $j = t + 1$ we see that the covariance between $x(t)$ and $x(t+1) - x(t+1|t+1)$ is

$$\mathrm{Cov}(x(t), x(t) - x(t|t))F^T(t)$$

$$-\mathrm{Cov}(x(t), \varepsilon(t+1))R^{-1}(t+1)H(t+1)S(t+1|t)$$

$$= S(t|t-1)M^T(t)[I$$

$$-H^T(t+1)R^{-1}(t+1)H(t+1)S(t+1|t)].$$

The conclusion then follows from post-multiplication by $F^T(t+1)H^T(t+2)$ and the definition of $M^T(t+1)$.

To establish the general case we proceed by induction and assume that for $k = t + j$

$$\mathrm{Cov}(x(t), \varepsilon(k-1))$$

$$= S(t|t-1)M^T(t) \cdots M^T(k-2)H^T(k-1).$$

$$(2.20)$$

As a result of (2.9) and (F2), this is equivalent to saying that the covariance between $x(t)$ and $x(k-1) - x(k-1|k-2)$ is

$$S(t|t-1)M^T(t) \cdots M^T(k-2). \qquad (2.21)$$

Then, using the same approach we employed for $j = t + 1, t + 2$, we apply (2.9), (F2), (2.13) and (F3) to establish that $\mathrm{Cov}(x(t), \varepsilon(k))$ is

$$\mathrm{Cov}(x(t), x(k-1) - x(k-1|k-1))F^T(k-1)H^T(k)$$

and thereby allow access to the induction hypothesis. For this purpose we again break $x(k-1|k-1)$ into components corresponding to the history prior to time index $k-1$ and the contribution from $\varepsilon(k-1)$. This strategy reveals that $\mathrm{Cov}(x(t), x(k-1) - x(k-1|k-1))$ is $\mathrm{Cov}(x(t), x(k-1) - x(k-1|k-2))$ minus

$$\mathrm{Cov}(x(t), \varepsilon(k-1))R^{-1}(k-1)\mathrm{Cov}(\varepsilon(k-1), x(k-1))$$

and the conclusion follows from (2.21)–(2.20) and (2.15).

∎

Returning now to Algorithms 2.2–2.3 it should be noted that we are only able to evaluate $\Sigma_{X\varepsilon}$ in order n^2 flops because of Algorithm 2.1 that returns $S(t|t-1)$, $R(t)$, $t=1,\ldots,n$, in $O(n)$ operations. In fact, Algorithm 2.3 is only feasible when all the $S(t|t-1)$ (and hence the $M(t) = F(t) - F(t)S(t|t-1)H^T(t)R^{-1}(t)H(t)$) have already been computed. That is, we can only use Algorithm 2.3 after a forward pass that incorporates the computations in Algorithm 2.1. One conclusion that can be drawn from this is that it makes more sense to combine Algorithms 2.1 and 2.2 into one unified recursion that also computes the $M(t)$ matrices for use in the backward covariance recursion.

Algorithm 2.4 This algorithm computes $S(t|t)$, $R(t)$, $S(t|t-1)$, $M(t)$, $t=1,\ldots,n$ and $\sigma_{X\varepsilon}(t,j), t=1,\ldots,n$, $j=1,\ldots,t$.

/*Initialization*/
$S(1|0) = F(0)S(0|0)F^T(0) + Q(0)$
$R(1) = H(1)S(1|0)H^T(1) + W(1)$
$S(1|1) = S(1|0) - S(1|0)H^T(1)R^{-1}(1)H(1)S(1|0)$
$M(1) = F(1) - F(1)S(1|0)H^T(1)R^{-1}(1)H(1)$
$\sigma_{X\varepsilon}(1,1) = S(1|0)H^T(1)$
for $t=2$ **to** n
    for $j=1$ **to** $(t-1)$
        $\sigma_{X\varepsilon}(t,j) = F(t-1)\sigma_{X\varepsilon}(t-1,j)$
    end for
    $S(t|t-1) = F(t-1)S(t-1|t-1)F^T(t-1)$
        $+Q(t-1)$
    $R(t) = H(t)S(t|t-1)H^T(t) + W(t)$
    $S(t|t) = S(t|t-1)$
        $-S(t|t-1)H^T(t)R^{-1}(t)H(t)S(t|t-1)$
    $M(t) = F(t) - F(t)S(t|t-1)H^T(t)R^{-1}(t)H(t)$
    $\sigma_{X\varepsilon}(t,t) = S(t|t-1)H^T(t)$
end for

Algorithm 2.4 returns everything that is needed to run the backward covariance recursion of Algorithm 2.3. This two-stage approach can be perfectly satisfactory and we will see this reflected in some of the forward and backward recursions for computing signal and state vector estimators in Chapters 4 and 5. However, there are also cases where it would be more convenient to have the entire matrix $\Sigma_{X\varepsilon}$ in hand after a single forward pass.

One way to evaluate the entirety of $\Sigma_{X\varepsilon}$ in a single recursion is to work forward from the upper left hand corner of the matrix in an L-shaped pattern. Computations proceed along the diagonal and below diagonal blocks for the tth row block using Algorithm 2.2. Then the above diagonal blocks for the $(t + 1)$st column block are evaluated. To see how this can be accomplished let us explicitly consider a few initial cases using Lemma 2.4.

The above diagonal blocks for row block 1 consist of the matrices $\sigma_{X\varepsilon}(1, 2) = S(1|0)M^T(1)H^T(2)$ and

$$\sigma_{X\varepsilon}(1, j) = S(1|0)M^T(1) \cdots M^T(j - 1)H^T(j)$$

for $j = 2, \ldots, n$. So, we could conceivably compute all the blocks in the first row block by recursively updating from $S(1|0)M^T(1)$ to $S(1|0)M^T(1)M^T(2)$, etc., followed by post-multiplication by the relevant $H^T(j)$ matrices to obtain the actual block element of $\Sigma_{X\varepsilon}$. The problem with this is that

$$M(t) = F(t) - F(t)S(t|t - 1)H^T(t)R^{-1}(t)H(t)$$

so that all the $M(t)$, $t = 2, \ldots, n$, will not be available unless we have already evaluated $S(t|t-1)$, $t = 1, \ldots, n$. Consequently, if we want to compute the $S(t|t - 1)$ and $R(t)$ in tandem with evaluation of $\Sigma_{X\varepsilon}$ we need a slightly more subtle strategy.

Now, in general, for the tth row block the above diagonal blocks appear like

$$\sigma_{X\varepsilon}(t, j) = S(t|t - 1)M^T(t) \cdots M^T(j - 1)H^T(j)$$

for $j = t+1, \ldots, n$. So, computations above the diagonal can be carried out by storing and updating matrices of the form

$$A(t, j) = S(t|t - 1)M^T(t) \cdots M^T(j - 1).$$

If $S(t|t - 1)$ is computed in conjunction with the diagonal and below diagonal blocks of the tth row block, then this allows for computation of $M(t)$ which, in turn, allows for computation of

$$\sigma_{X\varepsilon}(t, t + 1) = S(t|t - 1)M^T(t)H^T(t + 1)$$

$$= A(t, t + 1)H^T(t + 1).$$

In addition, it is now also possible to evaluate the other elements in the $(t + 1)$st column block since

$$\sigma_{X\varepsilon}(i, t + 1) = S(i|i - 1)M^T(i) \cdots M^T(t)H^T(t + 1)$$

$$= A(i, t + 1)H^T(t + 1)$$

for $i = t, \ldots, 1$ and

$$A(i, t + 1) = A(i, t)M^T(t)$$

provided that we take $A(t, t) := S(t|t - 1)$.

From a recursive standpoint the idea is to compute $S(t|t - 1)$ and $M(t)$ on the tth forward step which allows us (by updating) to obtain the matrices $A(k, t + 1)$ for $k = 1, \ldots, t$ that are needed to evaluate the above diagonal blocks for the $(t + 1)$st column block. For example, we can compute $A(1, 2) = A(1, 1)M^T(1)$ once we have $S(1|0)$ (and, hence, $M(1)$) and can evaluate $A(1, 3) = A(1, 2)M^T(2)$ and $A(2, 3) = A(2, 2)M^T(2)$ once we have $S(2|1)$ (and, hence, $M(2)$). Using these ideas we can develop a road map of sorts for moving forward across the above diagonal entries of $\Sigma_{X\varepsilon}$ that progresses from

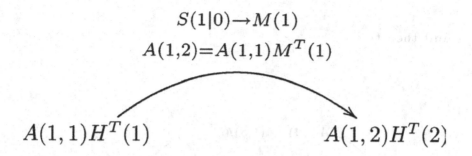

$$S(1|0) \to M(1)$$
$$A(1,2) = A(1,1)M^T(1)$$

$$A(1,1)H^T(1) \qquad\qquad A(1,2)H^T(2)$$

to

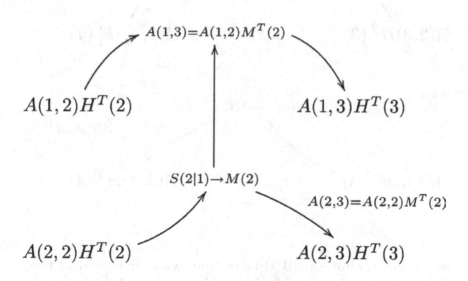

$$A(1,3) = A(1,2)M^T(2)$$

$$A(1,2)H^T(2) \qquad\qquad A(1,3)H^T(3)$$

$$S(2|1) \to M(2)$$

$$A(2,3) = A(2,2)M^T(2)$$

$$A(2,2)H^T(2) \qquad\qquad A(2,3)H^T(3)$$

and then to

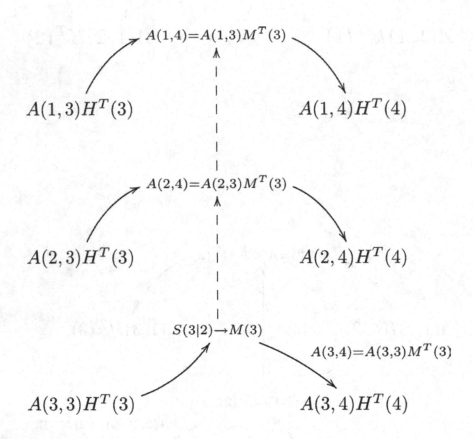

etc. The upshot of all this is that we can combine the below diagonal recursion from Algorithm 2.4 with a new recursion that computes above diagonal entries to evaluate the entire matrix $\Sigma_{X\varepsilon}$.

Algorithm 2.5 This algorithm computes $R(t)$, $S(t|t)$, $S(t|t-1)$, $t = 1, \ldots, n$ and $\sigma_{X\varepsilon}(t, j)$, $t, j = 1, \ldots, n$.

/*Initialization for forward recursion*/
$$S(1|0) = F(0)S(0|0)F^T(0) + Q(0)$$
$$R(1) = H(1)S(1|0)H^T(1) + W(1)$$
$$S(1|1) = S(1|0) - S(1|0)H^T(1)R^{-1}(1)H(1)S(1|0)$$
$$\sigma_{X\varepsilon}(1, 1) = S(1|0)H^T(1)$$
$$M(1) = F(1) - F(1)S(1|0)H^T(1)R^{-1}(1)H(1)$$
$$A(1) = S(1|0)M^T(1)$$
$$\sigma_{X\varepsilon}(1, 2) = A(1)H^T(2)$$
for $t = 2$ **to** n
    /*Computation of below diagonal blocks*/
    for $j = 1$ **to** $(t - 1)$
        $\sigma_{X\varepsilon}(t, j) = F(t - 1)\sigma_{X\varepsilon}(t - 1, j)$
    end for
    $S(t|t - 1) = F(t - 1)S(t - 1|t - 1)F^T(t - 1)$
        $+Q(t - 1)$
    $R(t) = H(t)S(t|t - 1)H^T(t) + W(t)$
    $S(t|t) = S(t|t - 1)$
        $-S(t|t - 1)H^T(t)R^{-1}(t)H(t)S(t|t - 1)$
    $\sigma_{X\varepsilon}(t, t) = S(t|t - 1)H^T(t)$
    /*Computation of above diagonal blocks*/
    If $t \leq n - 1$
        $M(t) = F(t)$
            $-F(t)S(t|t - 1)H^T(t)R^{-1}(t)H(t)$
        $A(t) = S(t|t - 1)M^T(t)$
        $\sigma_{X\varepsilon}(t, t + 1) = A(t)H^T(t + 1)$
        for $i = t - 1$ **to** 1
            $A(i) = A(i)M^T(t)$
            $\sigma_{X\varepsilon}(i, t + 1) = A(i)H^T(t + 1)$
        end for
    end if
end for

2.4   An example

To illustrate the results of the previous section consider
the state-space model where $H(t)$, $F(t)$, $Q(t)$ and $W(t)$
are time independent. In this case

$$y(t) = Hx(t) + e(t)$$

and

$$x(t + 1) = Fx(t) + u(t)$$

for known matrices H and F. In keeping with, e.g., the
ARMA example of Section 1.3 we also have $\text{Var}(e(t)) =$
W_0, $\text{Var}(u(t - 1)) = Q_0$ for $t = 1, \ldots, n$ and $S(0|0) = 0$
so that $x(0) = 0$. This formulation is also applicable to
the other example from Chapter 1 that involved sampling
from Brownian motion with white noise if the samples
are acquired at equidistant points. In this latter case we
have $p = q = 1$, $F = H = 1$ and Q_0 is the common dis-
tance $\tau_i - \tau_{i-1}$ between the points $\tau_1, \ldots, \tau_n$ at which
observations are taken.

Applying Lemma 2.4 we see that the below diagonal
blocks of $\Sigma_{X\varepsilon}$ have a relatively simple representation as

$$\sigma_{X\varepsilon}(t, j) = F^{t-j} S(j|j - 1) H^T, \ j \leq t - 1.$$

Expressions for the above diagonal entries are more complicated except in the case of univariate state and response variables.

To proceed further let us now specialize to the instance
where $p = q = 1$ so that H, F, W_0 and Q_0 are all scalar

valued. Then, from Lemmas 2.1–2.3 we obtain

$$S(t|t-1) = F^2 S(t-1|t-1) + Q_0, \qquad (2.22)$$

$$R(t) = H^2 S(t|t-1) + W_0, \qquad (2.23)$$

$$M(t) = F\left(1 - \frac{S(t|t-1)H^2}{R(t)}\right), \qquad (2.24)$$

$$S(t|t) = S(t|t-1)\left(1 - \frac{S(t|t-1)H^2}{R(t)}\right)$$

$$(2.25)$$

with initializing values provided by

$$S(0|0) = 0,$$

$$S(1|0) = F^2 S(0|0) + Q_0 = Q_0,$$

$$R(1) = H^2 Q_0 + W_0,$$

$$M(1) = F\left(1 - \frac{Q_0 H^2}{R(1)}\right)$$

and

$$S(1|1) = Q_0 - \frac{Q_0^2 H^2}{R(1)}.$$

The above diagonal entries are now somewhat more tractable and can be expressed as

$$\sigma_{X\varepsilon}(t,j) = F^{j-t} S(t|t-1) H \prod_{i=t}^{j-1}\left(1 - \frac{S(i|i-1)H^2}{R(i)}\right)$$

$$= \sigma_{X\varepsilon}(j,t) \prod_{i=t}^{j-1}\left(1 - \frac{S(i|i-1)H^2}{R(i)}\right),$$

for $j \geq t + 1$ which reveals a type of quasi-symmetry for the covariances.

Some further simplifications occur if we focus on large values of t. To see this, first use (2.23) to obtain

$$S(t|t - 1) = \frac{R(t) - W_0}{H^2} \qquad (2.26)$$

which, in conjunction with (2.25), produces

$$S(t|t) = \frac{W_0}{H^2} \left(1 - \frac{W_0}{R(t)} \right). \qquad (2.27)$$

Consequently,

$$R(t) = H^2 \left[F^2 \frac{W_0}{H^2} \left(1 - \frac{W_0}{R(t-1)} \right) + Q_0 \right] + W_0$$

$$= F^2 W_0 + H^2 Q_0 + W_0 - \frac{F^2 W_0^2}{R(t-1)}.$$

If we now let $C_1 = F^2 W_0 + H^2 Q_0 + W_0$ and $C_2 = F^2 W_0^2$, it then follows that

$$R(t) = C_1 - \frac{C_2}{R(t-1)}$$

$$= C_1 - \frac{C_2}{C_1 - \dfrac{C_2}{R(t-2)}}$$

$$= C_1 - \frac{C_2}{C_1 - \dfrac{C_2}{C_1 - \dfrac{C_2}{C_1 - \dfrac{C_2}{R(t-3)}}}}$$

which reveals a continued fraction representation for $R(t)$ as described, for example, in Khinchin (1997).

General results for convergence of continued fractions can be found in Chapter 3 of Wall (1948). In particular, one sufficient condition (Wall 1948, Theorem 10.1) for our setting is that $C_2/C_1^2 \leq 1/4$. This is true for our case because $C_2/C_1^2 \leq F^2/(1+F^2)^2$.

Denoting the limit of the $R(t)$ by $R(\infty)$, we can now approximate $R(t)$ by this limit in (2.26)–(2.27) and (2.24) to see that for large t

$$S(t|t-1) \approx \frac{R(\infty) - W_0}{H^2} \tag{2.28}$$

$$S(t|t) \approx \frac{W_0}{H^2} \left(1 - \frac{W_0}{R(\infty)} \right) \tag{2.29}$$

$$M(t) \approx \frac{F W_0}{R(\infty)} \tag{2.30}$$

and, hence,

$$\mathrm{Cov}(x(t), \varepsilon(t)) \approx \frac{R(\infty) - W_0}{H},$$

$$\mathrm{Cov}(x(t), \varepsilon(j)) \approx F^{t-j} \frac{R(\infty) - W_0}{H},$$

for $j \leq t-1$ and

$$\mathrm{Cov}(x(t), \varepsilon(j)) \approx F^{j-t} \frac{R(\infty) - W_0}{H} \left(\frac{W_0}{R(\infty)} \right)^{j-t}$$

for $j \geq t+1$.

Recall from Chapter 1 that the $R(t)$ have interpretations as the elements of the diagonal matrix R in the Cholesky decomposition $\mathrm{Var}(y) = LRL^T$ for the response variance-covariance matrix. Consequently, characterization of $R(\infty)$ provides us with information about how the elements of R (and, hence, the variances of the $\varepsilon(t)$) will behave for large t. To address this issue note from

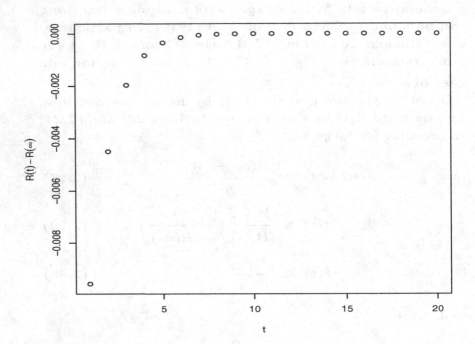

FIGURE 2.1

$R(t) - R(\infty)$ when $H = F = 1$, $W_0 = .05$, $Q_0 = .01$

the above analysis that $S(t|t-1)$ can be approximated by both $(R(\infty) - W_0)/H^2$ and

$$\frac{W_0 F^2}{H^2}\left(1 - \frac{W_0}{R(\infty)}\right) + Q_0.$$

By equating these two expressions one finds that $R(\infty)$ must satisfy the quadratic equation $r^2 - rC_1 + C_2 = 0$. The condition $C_1^2 \geq 4C_2$ is always satisfied in this case so that the two real roots of the equation are obtained from the quadratic formula as

$$\frac{C_1 \pm \sqrt{C_1^2 - 4C_2}}{2}.$$

The smaller root is strictly smaller than W_0 except when one of H, W_0 or Q_0 is zero. Thus, except for these exceptional cases, $R(\infty)$ is

$$\frac{F^2 W_0 + H^2 Q_0 + W_0}{2}$$

$$+ \ \frac{\sqrt{(F^2 W_0 + H^2 Q_0 + W_0)^2 - 4F^2 W_0^2}}{2} . \quad (2.31)$$

An illustration of the use of this approximation is provided by Figure 2.1 where the actual values of $R(t) - R(\infty)$ are shown for the case of sampling from Brownian Motion with white noise where $H = F = 1$ and we took $W_0 = .05$ and $Q_0 = .01$ to obtain $R(\infty) = .0779$.

3

Recursions for L and L^{-1}

3.1 Introduction

In this chapter we will develop algorithms for computing the matrices L and L^{-1} that arise in the Cholesky factorization of $\text{Var}(y)$, where y is the vector of responses from a state-space model. There are several reasons for considering this particular problem.

Our approach to the signal-plus-noise prediction problem revolves around the innovation vectors that are computed by applying the Gramm-Schmidt orthogonalization method to the response vectors. In Section 1.2.3 we defined innovations recursively by starting with $\varepsilon(1) = y(1)$ and then successively computing

$$\varepsilon(t) = y(t) - \sum_{j=1}^{t-1} L(t, j)\varepsilon(j), \qquad (3.1)$$

for $t = 2, \ldots, n$, with $L(t, j)$ being the matrix in the jth block column of the tth block row of the lower triangular matrix L in the Cholesky decomposition

$$\text{Var}(y) = LRL^T. \qquad (3.2)$$

Thus, computation of the innovations is intimately linked to the evaluation of L.

53

From another perspective, we saw in Chapter 1 that
(3.1) arises from the forward substitution step for solving
the lower triangular, linear equation system $L\varepsilon = y$: that
is,

$$\varepsilon = L^{-1}y.$$

This, in turn, was seen to have the consequence that the
BLUP of f based on y had the form

$$\hat{f} = y - W(L^T)^{-1}R^{-1}\varepsilon$$

with prediction error variance-covariance matrix

$$V = W - W(L^T)^{-1}R^{-1}L^{-1}W.$$

Thus, at least implicitly, the matrix L^{-1} plays a role in
both the prediction of the signal as well as assessment
of the accuracy of the predictor.

In the next section we study efficient methods for computing L. In this regard, we develop a forward recursion
that produces the matrix row by row starting from its
upper left block entry. Section 3.3 then provides a parallel result pertaining to L^{-1}.

3.2   Recursions for L

For the developments in this and subsequent sections it
will be convenient to introduce a final piece of notation
for the so-called *Kalman gain matrices*. These matrices arise
naturally in formulae for both L and L^{-1} and, not surprisingly, appear in various signal and state vector prediction formulae that we will encounter in the next chapter. They are defined by

$$K(t) = F(t)S(t|t - 1)H^T(t)R^{-1}(t), \ t = 1, \ldots, n.$$

$$(3.3)$$

The Kalman gain matrices involve a number of factors that we have already experienced in Chapter 2 and, in particular, are intimately related to the ubiquitous matrices (2.17) via the relation $M(t) = F(t) - K(t)H(t)$. With this notational preliminary, we can now give a detailed description of the form of L.

Theorem 3.1 *Let* $L(t, j)$, $t, j = 1, \ldots, n$, *be* $p \times p$ *matrices with* $L = \{L(t, j)\}_{t,j=1:n}$ *for L in (3.2). Then,* $L(t, t) = I$, $L(t, j) = 0$ *for* $j > t$,

$$L(t, t - 1)$$
$$= H(t)F(t - 1)S(t - 1|t - 2)H^T(t - 1)R^{-1}(t - 1)$$
$$= H(t)K(t - 1) \tag{3.4}$$

for $t = 2, \ldots, n$, *and*

$$L(t, j) = H(t)F(t - 1) \cdots F(j)S(j|j - 1)H^T(j)R^{-1}(j)$$
$$= H(t)F(t - 1) \cdots F(j + 1)K(j) \tag{3.5}$$

for $t = 3, \ldots, n$ *and* $j = 1, \ldots, t - 2$.

Proof. From (1.16) we have for $t > j$ that

$$L(t, j) = \text{Cov}(y(t), \varepsilon(j))R^{-1}(j).$$

But, $y(t) = H(t)x(t) + e(t)$ and $e(t)$ is uncorrelated with $\varepsilon(1), \ldots, \varepsilon(t - 1)$. Thus,

$$L(t, j) = H(t) \text{Cov}(x(t), \varepsilon(j))R^{-1}(j)$$

and the result is a consequence of (2.16) in Lemma 2.4. ∎

Efficient order n^2 (for p, q small relative to n) recursions for L can now be obtained in several ways. One approach is to build L row by row. To see how this can be accomplished, use Theorem 3.1 to see that the first column for L is

$$I$$
$$H(2)F(1)S(1|0)H^T(1)R^{-1}(1)$$
$$H(3)F(2)F(1)S(1|0)H^T(1)R^{-1}(1)$$
$$H(4)F(3)F(2)F(1)S(1|0)H^T(1)R^{-1}(1)$$
$$H(5)F(4)\cdots F(1)S(1|0)H^T(1)R^{-1}(1)$$

$$\vdots$$

$$H(n-1)F(n-2)\cdots F(1)S(1|0)H^T(1)R^{-1}(1)$$
$$H(n)F(n-1)\cdots F(1)S(1|0)H^T(1)R^{-1}(1)$$

while the second has the form

$$0$$
$$I$$
$$H(3)F(2)S(2|1)H^T(2)R^{-1}(2)$$
$$H(4)F(3)F(2)S(2|1)H^T(2)R^{-1}(2)$$
$$H(5)F(4)F(3)F(2)S(2|1)H^T(2)R^{-1}(2)$$

$$\vdots$$

$$H(n-1)F(n-2)\cdots F(2)S(2|1)H^T(2)R^{-1}(2)$$
$$H(n)F(n-1)\cdots F(2)S(2|1)H^T(2)R^{-1}(2)$$

and the third is

$$0$$
$$0$$
$$I$$
$$H(4)F(3)S(3|2)H^T(3)R^{-1}(3)$$
$$H(5)F(4)F(3)S(3|2)H^T(3)R^{-1}(3)$$

$$\vdots$$

$$H(n-1)F(n-2)\cdots F(3)S(3|2)H^T(3)R^{-1}(3)$$
$$H(n)F(n-1)\cdots F(3)S(3|2)H^T(3)R^{-1}(3)$$

Some of the structures in L becomes apparent from examining these special cases: namely, the rows for each column can be built up through successive evaluation of the matrices

$$A(t, j) = F(t-1) \cdots F(j)S(j|j-1)H^T(j)R^{-1}(j)$$

$$(3.6)$$

for $t = j+1, \ldots, n$ since $L(t, j) = H(t)A(t, j)$.

Formula (3.6) suggests several ways to approach the recursive evaluation of L. For example, we could start in the upper left hand corner of L, compute $S(1|0)$ and $R(1)$ and then fill out the below diagonal elements for the first block column. Then $S(1|0)$, $R(1)$ can be updated to $S(2|1)$, $R(2)$ using Algorithm 2.1 and the elements of the second block column could be obtained, etc. An alternative approach is to fill out L on a row (block) by row (block) basis.

To be a bit more specific let us consider how to obtain the $(t+1)$st block row after we have computed the tth block row. With this in mind, observe that the first $t-1$ column blocks in the $(t+1)$st row are

$$H(t+1)\Big[A(t+1, 1)\ \ A(t+1, 2)\ \ \ldots\ \ A(t+1, t-1) \Big]$$

$$= H(t+1)F(t)\Big[A(t, 1)\ \ A(t, 2)\ \ \ldots\ \ A(t, t-1) \Big]$$

as a result of the update formula

$$A(t+1, j) = F(t)A(t, j), \quad j = 1, \ldots, t-1.$$

This idea produces the following algorithm.

Algorithm 3.1 This algorithm evaluates L row by row beginning with the upper left hand row block.

$$L(1, 1) = I$$
$$A(2, 1) = F(1)S(1|0)H^T(1)R^{-1}(1)$$

$L(2, 1) = H(2)A(2, 1)$
for $t = 3$ **to** n
    $L(t, t) = I$
    $A(t, t - 1) =$
        $F(t - 1)S(t - 1|t - 2)H^T(t - 1)R^{-1}(t - 1)$
    $L(t, t - 1) = H(t)A(t, t - 1)$
    for $j = 1$ **to** $t - 2$
        $A(t, j) = F(t - 1)A(t - 1, j)$
        $L(t, j) = H(t)A(t, j)$
    end for
end for

There is nothing special about the forward approach to computing L. In fact, it is also possible to compute the block matrices that comprise L in reverse order by working backward from the lower right hand block of L to its upper left corner. Identical comments apply to the algorithm developed for L^{-1} in the next section. Our primary reason for concentrating on forward recursions here is because the resulting algorithms are more closely related to the classical filtering recursions for computing the innovations and signal and state vector predictors that we will study in the next chapter.

3.3   Recursions for L^{-1}

In this section we develop a parallel of Algorithm 3.1 for use in the evaluation of L^{-1} rather than L. We begin by establishing an analog of Theorem 3.1 for the inverse matrix.

Theorem 3.2   *The matrix* $L^{-1} = \{L^{-1}(t, j)\}_{t,j=1:n}$ *is lower triangular with identity matrix diagonal blocks. For* $t =$

2, . . ., n,

$$L^{-1}(t, t-1) = -H(t)K(t-1) \qquad (3.7)$$

and

$$L^{-1}(t, j) = -H(t)M(t-1) \cdots M(j+1)K(j)$$

$$(3.8)$$

for j = 1, . . ., t − 2 and t = 3, . . ., n.

Proof. To prove this result we will employ Theorem A.1 from the Appendix which gives the result that $L^{-1}(t, t-1) = -L(t, t-1)$ as well as the relation

$$L^{-1}(t, j) = -L(t, j) - \sum_{i=j+1}^{t-1} L(t, i)L^{-1}(i, j) \quad (3.9)$$

for $j = 1, . . ., t − 2$. Thus, (3.7) is immediate from (3.4). Now, for $j = t − 2$ and $t = 3, . . ., n$, (3.9) reveals that

$$L^{-1}(t, t-2) = -H(t)F(t-1)K(t-2)$$

$$-L(t, t-1)L^{-1}(t-1, t-2)$$

$$= -H(t)F(t-1)K(t-2)$$

$$+H(t)K(t-1)H(t-1)K(t-2)$$

$$= -H(t)M(t-1)K(t-2) \qquad (3.10)$$

due to (3.4), (3.5) and (3.7). Next, for $t = 4, . . ., n$, the application of (3.9) produces

$$L^{-1}(t, t - 3) = -H(t)F(t - 1)F(t - 2)K(t - 3)$$

$$-L(t, t - 2)L^{-1}(t - 2, t - 3)$$

$$-L(t, t - 1)L^{-1}(t - 1, t - 3)$$

$$= -H(t)F(t - 1)F(t - 2)K(t - 3)$$

$$+H(t)F(t - 1)K(t - 2)H(t - 2)K(t - 3)$$

$$+H(t)K(t - 1)H(t - 1)M(t - 2)K(t - 3)$$

$$= -H(t)F(t - 1)\Big[F(t - 2)$$

$$-K(t - 2)H(t - 2)\Big]K(t - 3)$$

$$+H(t)K(t - 1)H(t - 1)M(t - 2)K(t - 3)$$

$$= -H(t)M(t - 1)M(t - 2)K(t - 3) \qquad (3.11)$$

using (3.4), (3.5), (3.7) and (3.10).

At this point we have established that Theorem 3.2 is valid for the sub-diagonal blocks, sub-sub-diagonal blocks and the sub-sub-sub diagonal blocks of L^{-1} as a result of showing (3.7), (3.10) and (3.11), respectively. We will continue in this manner by proving that (3.8) holds along the kth sub-diagonal of L^{-1}. This means we want to verify (3.8) for the matrices $L^{-1}(k + 1, 1), \ldots, L^{-1}(n, n - k)$. The proof will be completed once we show this is true for the as yet untreated cases of $k = 4, \ldots, n - 1$ which correspond to all the remaining blocks of L^{-1}.

We now wish to prove that

$$L^{-1}(t, t - k)$$

$$= -H(t)M(t - 1) \cdots M(t - k + 1)K(t - k)$$

for $t = k + 1, \ldots, n$ and any $4 \le k \le n - 1$. To simplify matters slightly, observe that for such cases we have

$$L^{-1}(t, t - k) = H(t)A(t, t - k)$$

with

$A(t, t - k)$

$\quad = -F(t - 1) \cdots F(t - k + 1)K(t - k)$

$$\quad\quad - \sum_{i=t-k+1}^{t-2} F(t - 1) \cdots F(i + 1)K(i)L^{-1}(i, t - k)$$

$$\quad\quad - K(t - 1)L^{-1}(t - 1, t - k) \tag{3.12}$$

as a result of (3.9) and (3.4)–(3.5). Thus, it suffices to show that

$$A(t, t - k) = -M(t - 1) \cdots M(t - k + 1)K(t - k) \tag{3.13}$$

for $t = k + 1, \ldots, n$.

Proceeding now by induction, suppose that (3.13) holds for some integer $k \geq 1$. Then,

$A(t, t - k - 1)$

$\quad = -F(t - 1) \cdots F(t - k)K(t - k - 1)$

$$\quad\quad - \sum_{i=t-k}^{t-2} F(t - 1) \cdots F(i + 1)K(i)L^{-1}(i, t - k - 1)$$

$$\quad\quad - K(t - 1)L^{-1}(t - 1, t - k - 1)$$

$$\quad = F(t - 1)\Big[-F(t - 2) \cdots F(t - k)K(t - k - 1)$$

$$\quad\quad - \sum_{i=t-k}^{t-3} F(t - 2) \cdots F(i + 1)K(i)L^{-1}(i, t - k - 1)$$

$$\quad\quad - K(t - 2)L^{-1}(t - 2, t - k - 1)\Big]$$

$$\quad\quad - K(t - 1)L^{-1}(t - 1, t - k - 1)$$

$$\quad = F(t - 1)A(t - 1, t - k - 1)$$

$$\quad\quad - K(t - 1)H(t - 1)A(t - 1, t - k - 1).$$

But, this last expression is just $M(t-1)A(t-1, t-k-1)$ which proves the theorem. ∎

Let us now use Theorem 3.2 to construct an efficient scheme for computing L^{-1}. From (3.7)–(3.8) we see that the first column for L^{-1} is

$$I$$
$$-H(2)K(1)$$
$$-H(3)M(2)K(1)$$
$$-H(4)M(3)M(2)K(1)$$
$$-H(5)M(4)M(3)M(2)K(1)$$

$$\vdots$$

$$-H(n-1)M(n-2)\cdots M(2)K(1)$$
$$-H(n)M(n-1)\cdots M(2)K(1)$$

while the second has the form

$$0$$
$$I$$
$$-H(3)K(2)$$
$$-H(4)M(3)K(2)$$
$$-H(5)M(4)M(3)K(2)$$

$$\vdots$$

$$-H(n-1)M(n-2)\cdots M(3)K(2)$$
$$-H(n)M(n-1)\cdots M(3)K(2)$$

and the third is

$$0$$
$$0$$
$$I$$
$$-H(4)K(3)$$
$$-H(5)M(4)K(3)$$

$$\vdots$$

$$-H(n-1)M(n-2)\cdots M(4)K(3)$$
$$-H(n)M(n-1)\cdots M(4)K(3)$$

Thus, it appears that an almost identical strategy to the one we employed for computing L can be used for evaluating L^{-1} provided we replace the matrices (3.6) in the recursion by

$$A(t, j) = M(t - 1) \cdots M(j + 1)K(j). \qquad (3.14)$$

Algorithm 3.2 provides the resulting row by row forward recursion.

Algorithm 3.2 This algorithm evaluates L^{-1} row by row beginning with the upper left hand row block.

$$L^{-1}(1, 1) = I$$
$$A(2, 1) = K(1)$$
$$L^{-1}(2, 1) = -H(2)A(2, 1)$$
for $t = 3$ **to** n
    $L^{-1}(t, t) = I$
    $A(t, t - 1) = K(t - 1)$
    $L^{-1}(t, t - 1) = -H(t)A(t, t - 1)$
    for $j = 1$ **to** $t - 2$
        $A(t, j) = M(t - 1)A(t - 1, j)$
        $L^{-1}(t, j) = -H(t)A(t, j)$
    end for
end for

When evaluating the tth row block of L^{-1} using Algorithm 3.2 we need access to the matrices $K(j), M(j), j = 1, \ldots, t - 1$. These matrices are, in turn, simple functions of the matrices $R(j), S(j|j - 1), j = 1, \ldots, t - 1$, and Algorithm 2.1 evaluates these quantities in a way that is ideally suited for use in the forward computation of L^{-1}. By combining these two recursions one can thereby obtain an order n^2 algorithm that computes all the nonzero element of L^{-1} in one forward pass.

3.4   An example

To illustrate some of the ideas developed in this chapter
let us again consider the special case of the state-space
model discussed in Section 2.4. In this instance we have

$$y(t) = Hx(t) + e(t) \tag{3.15}$$

and

$$x(t + 1) = Fx(t) + u(t) \tag{3.16}$$

for known matrices H and F that do not depend on t.
The variance-covariance matrices for the e and u processes are also time invariant in that there are positive
definite matrices Q_0 and W_0 such that $Q(t-1) = Q_0$
and $W(t) = W_0$ for $t = 1, \ldots, n$.

An application of Theorem 3.1 reveals that when (3.15)–
(3.16) hold we will have

$$L = \begin{bmatrix} I & 0 & \cdots & 0 & 0 \\ HK(1) & I & \cdots & 0 & 0 \\ HFK(1) & HK(2) & \cdots & 0 & 0 \\ \vdots & \vdots & \ddots & \vdots & \vdots \\ HF^{n-2}K(1) & HF^{n-3}K(2) & \cdots & HK(n-1) & I \end{bmatrix}.$$

This particular example makes it easy to see that in some
respects the matrix L is composed of only the $n-1$ u-
nique factors $K(1), \ldots, K(n-1)$. Of course, this truly
is the case when $H = I$ and $F = I$. A similar development for L^{-1} using Theorem 3.2 shows that the below
diagonal blocks in the jth column of L^{-1} have the form

$$L^{-1}(j + 1, j) = -HK(j)$$

and

$$L^{-1}(t, j) = -HM(t - 1) \cdots M(j + 1)K(j)$$

for $t = j + 2, \ldots, n$.

To simplify matters a bit take $p = q = 1$ and consider the case where t is large so that the approximations (2.28)–(2.30) from Section 2.4 can be utilized. In this instance, (2.28) has the consequence that

$$K(t) = FS(t|t - 1)HR^{-1}(t)$$

$$\approx \frac{F}{H}\left(1 - \frac{W_0}{R(\infty)}\right).$$

Combining this result with our formulas for the elements of L gives

$$L(t, t - 1) \approx F\left(1 - \frac{W_0}{R(\infty)}\right),$$

when t is large, and

$$L(t, j) \approx F^{t-j}\left(1 - \frac{W_0}{R(\infty)}\right).$$

Since $M(t) = F - K(t)H \approx FW_0/R(\infty)$, parallel formulae for the elements of L^{-1} include

$$L^{-1}(t, t - 1) \approx -F\left(1 - \frac{W_0}{R(\infty)}\right)$$

which applies to cases where t is large. If we assume that j is sufficiently large that (2.28) can be applied to $M(j + 1)$ and $K(j)$, then we also have

$$L^{-1}(t, j) \approx \left(\frac{FW_0}{R(\infty)}\right)^{t-j}\left(1 - \frac{R(\infty)}{W_0}\right).$$

In the special case that $|F| \leq 1$, our approximations suggest that L^{-1} is diagonally dominant in the sense that $L^{-1}(t, j)$ decays to zero as j decreases. This is because $W_0/R(\infty) < 1$ if H, W_0 and Q_0 are nonzero.

4

Forward Recursions

4.1 Introduction

In this chapter we develop the standard Kalman filter (abbreviated as KF hereafter) forward recursions for prediction of the signal and state vectors. The basic *filtering* premise is that we are observing our data or response vectors in a sequence corresponding to the "time" index t. So, we first see $y(1)$, then $y(2)$, etc. At any given point in time t we have observed $y(1), \ldots, y(t)$ from the state-space model (1.19)–(1.28) and want to use this data to predict the values of the state vector $x(t)$ and corresponding signal vector $f(t) = H(t)x(t)$.

To accomplish the prediction we follow the plan laid out in Chapter 1. First we translate the response vectors $y(1), \ldots, y(t)$ to the innovation vectors $\varepsilon(1), \ldots, \varepsilon(t)$ using the Gramm-Schmidt method. Then, the BLUP of $x(t)$ based on $y(1), \ldots, y(t)$ is

$$x(t|t) = \sum_{j=1}^{t} \text{Cov}(x(t), \varepsilon(j)) R^{-1}(j)\varepsilon(j) \qquad (4.1)$$

and from Theorem 1.1 the BLUP of $f(t)$ is $H(t)x(t|t)$ or, equivalently,

$$f(t|t) = \sum_{j=1}^{t} \text{Cov}(f(t), \varepsilon(j)) R^{-1}(j)\varepsilon(j). \qquad (4.2)$$

In the next section we show how to efficiently compute the innovation vectors using the results of the previous chapter on the form of L and L^{-1}. Then, in Section 4.4 we employ Lemma 2.4 to derive algorithms for prediction of $x(t)$ and $f(t)$. Section 4.3 discusses another forward recursive scheme that updates previous predictions to account for the presence of new responses. Finally, Section 4.5 considers two examples: namely, the case of state-space processes where the matrices $H(t)$, $F(t)$, $W(t)$ and $Q(t)$ are all time invariant and the case of sampling from Brownian motion with white noise.

4.2   Computing the innovations

Once again write the Cholesky factorization of $\operatorname{Var}(y)$ as

$$\operatorname{Var}(y) = L R L^T,$$

where $L = \{L(t,j)\}_{t,j=1:n}$ is an $np \times np$ block lower triangular matrix having identity matrix diagonal blocks and R is a block diagonal matrix with diagonal blocks $R(1), \ldots, R(n)$. Then, from our developments in Sections 1.2.2–1.2.3 we know a number of things about the innovation process and its relationship to the Cholesky factorization. Most pertinent at this point is the fact that by combining (1.7)–(1.10) we can obtain

$$\varepsilon(1) = y(1), \tag{4.3}$$

$$R(1) = \operatorname{Var}(\varepsilon(1)) = \operatorname{Var}(y(1)) \tag{4.4}$$

and, for $t = 2, \ldots, n$,

$$\varepsilon(t) = y(t) - \sum_{j=1}^{t-1} \mathrm{Cov}(y(t), \varepsilon(j)) R^{-1}(j) \varepsilon(j)$$

$$= y(t) - \sum_{j=1}^{t-1} L(t, j) \varepsilon(j), \qquad (4.5)$$

with

$$R(t) = \mathrm{Var}(\varepsilon(t)). \qquad (4.6)$$

We have also previously noted that formulae (4.3)–(4.6) are the result of forward solution of the system $L\varepsilon = y$ which means that $\varepsilon = L^{-1}y$.

Given all the information we have collected about the form of L and L^{-1} in the previous chapter, it now seems natural to attempt to apply this knowledge to the development of computational methodology that can be used to obtain the innovation vectors. The next theorem is what results from this approach.

Theorem 4.1 *The innovation vectors satisfy*

$$\varepsilon(1) = y(1), \qquad (4.7)$$

$$\varepsilon(2) = y(2) - H(2)K(1)\varepsilon(1) \qquad (4.8)$$

$$= y(2) - H(2)K(1)y(1) \qquad (4.9)$$

and, for $t = 3, \ldots, n,$

$$\varepsilon(t) = y(t) - H(t)K(t-1)\varepsilon(t-1)$$

$$-H(t)\sum_{j=1}^{t-2} F(t-1)\cdots F(j+1)K(j)\varepsilon(j)$$

$$(4.10)$$

$$= y(t) - H(t)K(t-1)y(t-1)$$

$$-H(t)\sum_{j=1}^{t-2} M(t-1)\cdots M(j+1)K(j)y(j).$$

$$(4.11)$$

Proof. Formulas (4.7)–(4.11) are all obtained by straightforward applications of Theorems 3.1 and 3.2. Relations (4.7), (4.8) and (4.10) derive from using (4.3)–(4.6) in conjunction with (3.4) and (3.5). In contrast, (4.9) and (4.11) are a consequence of the fact that $\varepsilon = L^{-1}y$ and (3.7)–(3.8). ∎

Theorem 4.1 provides us with two ways to iteratively compute the innovation vectors. The first is a result of (4.7)–(4.8) and (4.10) that works only with the innovation vectors. The idea is that one can accumulate the sum

$$A(t-1) = K(t-1)\varepsilon(t-1)$$

$$+ \sum_{j=1}^{t-2} F(t-1)\cdots F(j+1)K(j)\varepsilon(j)$$

$$(4.12)$$

using the simple update relation

$$A(t) = K(t)\varepsilon(t) + F(t)A(t-1). \qquad (4.13)$$

Then, $\varepsilon(t+1) = y(t+1) - H(t+1)A(t)$ produces the next innovation vector in the sequence. We spell this out in Algorithm 4.1.

Algorithm 4.1 This algorithm evaluates $\varepsilon(1), \ldots, \varepsilon(n)$ in sequence using only the innovation vectors computed on previous steps.

$$\varepsilon(1) = y(1)$$
$$A(1) = K(1)\varepsilon(1)$$
for $t = 2$ **to** $n - 1$
$$\varepsilon(t) = y(t) - H(t)A(t-1)$$
$$A(t) = K(t)\varepsilon(t) + F(t)A(t-1)$$
end for
$$\varepsilon(n) = y(n) - H(n)A(n-1)$$

Perhaps the most startling feature of this algorithm is how little computational labor it involves. A brute force application of formula (4.10) to the computation of $\varepsilon(t)$ that involved repeated re-computation and re-summing of the elements in this expression would necessarily require an overall effort of order n^3. However, because we only need the sum, rather than each individual term in the sum, the updating formula (4.13) allows us to reduce the number of calculations on each step to two (matrix) multiplications and an addition. Consequently, the entire vector ε is obtained in $O(n)$ operations. Since the combined innovation vector $\varepsilon = (\varepsilon^T(1), \ldots, \varepsilon^T(n))^T$ consists of n blocks in its own right, this is the best possible result in terms of the total order of calculations. We should also note that there is no trickery involved here because the Kalman gain matrices that are required at each step of the algorithm can be computed concurrently, also in order n operations, using Algorithm 2.1.

Whereas Algorithm 4.1 computes the innovations using the innovations themselves, it is also possible to compute the innovations directly from the response vectors. The key to this result is (4.11) which has the implication that we can proceed as we did for Algorithm 4.1 except that now we accumulate on

$$A(t-1) = K(t-1)y(t-1)$$

$$+ \sum_{j=1}^{t-2} M(t-1) \cdots M(j+1)K(j)y(j)$$

which can be updated via the relation

$$A(t) = K(t)y(t) + M(t)A(t-1)$$

with $\varepsilon(t+1) = y(t+1) - H(t+1)A(t)$ as before. Algorithm 4.2 that results from this also returns ε in a total of order n operations.

Algorithm 4.2 This algorithm evaluates $\varepsilon(1), \ldots, \varepsilon(n)$ in sequence by direct use of the response vectors.

$$\varepsilon(1) = y(1)$$
$$A(1) = K(1)\varepsilon(1)$$
for $t = 2$ **to** $n - 1$
$$\quad \varepsilon(t) = y(t) - H(t)A(t-1)$$
$$\quad A(t) = K(t)y(t) + M(t)A(t-1)$$
end for
$$\varepsilon(n) = y(n) - H(n)A(n-1)$$

4.3   State and signal prediction

In this section we derive expressions for the BLUPs of the state and signal vectors. We have actually already encountered one such BLUP in the previous section where we computed the innovations, although we did not explicitly identify it as such at that time. We will now clarify this connection and then proceed to develop efficient recursions for computing signal and state vector predictors using the innovations.

The component blocks of the innovation vector are related to the one step ahead BLUPs $x(2|1), \ldots, x(n|n-1)$. To see this recall that under the state-space modeling framework

$$y(t) = H(t)x(t) + e(t)$$

with $e(t)$ being uncorrelated with $\varepsilon(1), \ldots, \varepsilon(t-1)$. Thus,

$$\varepsilon(t) = y(t) - \sum_{j=1}^{t-1} \text{Cov}(y(t), \varepsilon(j))R^{-1}(j)\varepsilon(j)$$

$$= y(t) - \sum_{j=1}^{t-1} H(t)\text{Cov}(x(t), \varepsilon(j))R^{-1}(j)\varepsilon(j)$$

$$= y(t) - H(t)x(t|t-1). \tag{4.14}$$

Consequently, Theorem 4.1 has the consequence that

$$x(t|t-1) = K(t-1)\varepsilon(t-1)$$

$$+ \sum_{j=1}^{t-2} F(t-1)\cdots F(j+1)K(j)\varepsilon(j)$$

$$\tag{4.15}$$

and we realize that $x(t|t-1)$ is identical to the vector $A(t-1)$ in (4.12) that was used in Algorithm 4.1. Alternatively, expression (4.15) can be obtained directly from (2.16) in Lemma 2.4. In any case, (4.14) has the practical implication that we can compute $x(t|t-1)$ in tandem with $\varepsilon(t)$ or conversely.

Now note that

$$x(t|t) = \sum_{j=1}^{t} \text{Cov}(x(t), \varepsilon(j))R^{-1}(j)\varepsilon(j)$$

$$= \text{Cov}(x(t), \varepsilon(t))R^{-1}(t)\varepsilon(t) + x(t|t-1)$$

$$\tag{4.16}$$

and that Lemma 2.4 gives

$$\text{Cov}(x(t), \varepsilon(t)) = S(t|t-1)H^T(t). \qquad (4.17)$$

Equations (4.15)–(4.17) produce useful expressions for the BLUPs $x(t|t-1)$ and $x(t|t)$ that we now collect and summarize in Theorem 4.2.

Theorem 4.2 *The BLUP of $x(1)$ based on $y(1)$ is*

$$x(1|1) = S(1|0)H^T(1)R^{-1}(1)\varepsilon(1). \qquad (4.18)$$

For $t = 2, \ldots, n$ the BLUP of $x(t)$ based on $y(1), \ldots, y(t-1)$ is

$$x(t|t-1) = F(t-1)x(t-1|t-1) \qquad (4.19)$$

and the BLUP of $x(t)$ based on $y(1), \ldots, y(t)$ is

$$x(t|t) = S(t|t-1)H^T(t)R^{-1}(t)\varepsilon(t) + x(t|t-1). \quad (4.20)$$

Proof. Result (4.20) is immediate from (4.16)–(4.17). Verification of (4.19) can be accomplished through an application of Proposition 1.1. Alternatively, for a purely algebraic approach, use (4.15)–(4.17) and the definition of the Kalman gain matrices in (3.3) to see that

$$
\begin{aligned}
x(2|1) &= F(1)S(1|0)H^T(1)R^{-1}(1)\varepsilon(1) \\
&= F(1)x(1|1), \\
x(3|2) &= F(2)S(2|1)H^T(2)R^{-1}(2)\varepsilon(2) \\
&\quad + F(2)K(1)\varepsilon(1) \\
&= F(2)x(2|2)
\end{aligned}
$$

and, for $t > 3$, $x(t|t - 1)$ has the general form

$$F(t - 1)S(t - 1|t - 2)H^T(t - 1)R^{-1}(t - 1)\varepsilon(t - 1)$$

$$+ F(t - 1)K(t - 2)\varepsilon(t - 2)$$

$$+ F(t - 1)\sum_{j=1}^{t-3} F(t - 2) \cdots F(j + 1)K(j)\varepsilon(j)$$

$$= F(t - 1)x(t - 1|t - 1).$$

■

The intimate connection between $x(t|t - 1)$ and $\varepsilon(t)$ revealed in (4.14) means that we can compute the innovations and state and signal vector predictors all in one combined forward recursion. We give a complete treatment of the resulting classic, order n, KF algorithm below that includes the recursive evaluation of all the quantities that are required to produce the predictors. Note that in carrying out the recursions we are using the result from Theorem 1.1 that $f(t|t - 1) = H(t)x(t|t - 1)$, $f(t|t) = H(t)x(t|t)$, $V(t|t - 1) = H(t)S(t|t - 1)H^T(t)$ and $V(t|t) = H(t)S(t|t)H^T(t)$ to return signal estimators and their prediction error variance-covariance matrices along with the parallel quantities for state vector prediction.

Algorithm 4.3 This algorithm returns the predictors $x(t|t - 1)$, $x(t|t)$, $f(t|t - 1)$, $f(t|t)$, $t = 1, \ldots, n$, along with their associated prediction error variance-covariance matrices $S(t|t-1)$, $S(t|t)$, $V(t|t-1)$, $V(t|t)$, $t = 1, \ldots, n$.

/*Initialization*/
$S(1|0) = F(0)S(0|0)F^T(0) + Q(0)$
$R(1) = H(1)S(1|0)H^T(1) + W(1)$
$S(1|1) = S(1|0) - S(1|0)H^T(1)R^{-1}(1)H(1)S(1|0)$
$\varepsilon(1) = y(1)$
$x(1|1) = S(1|0)H^T(1)R^{-1}(1)\varepsilon(1)$

for $t = 2$ **to** n

$\quad S(t|t-1) = F(t-1)S(t-1|t-1)F^T(t-1)$
$\quad\quad + Q(t-1)$
$\quad V(t|t-1) = H(t)S(t|t-1)H^T(t)$
$\quad R(t) = H(t)S(t|t-1)H^T(t) + W(t)$
$\quad S(t|t) = S(t|t-1)$
$\quad\quad -S(t|t-1)H^T(t)R^{-1}(t)H(t)S(t|t-1)$
$\quad V(t|t) = H(t)S(t|t)H^T(t)$
$\quad x(t|t-1) = F(t-1)x(t-1|t-1)$
$\quad f(t|t-1) = H(t)x(t|t-1)$
$\quad \varepsilon(t) = y(t) - H(t)x(t|t-1)$
$\quad x(t|t) = S(t|t-1)H^T(t)R^{-1}(t)\varepsilon(t)$
$\quad\quad + x(t|t-1)$
$\quad f(t|t) = H(t)x(t|t)$

end for

The term *filter* that is used to describe this algorithm has at least two connotations. The first corresponds to the idea of predicting or "recovering" the signal f by "filtering out" the inherent noise (i.e., $e(1), \ldots, e(n)$) from the data. The phrase filter is also used to designate a particular type of data processing for when data arrives in sequence with respect to the "time" variable t. In this instance filtering also indicates that one is attempting to predict the signal $f(t)$ using only the responses $y(1), \ldots, y(t)$ that have become available at that point in time.

4.4   Other options

The KF Algorithm 4.3 returns the state vector predictors $x(1|1)$, $x(2|2)$, $x(3|3)$, etc. However, when we compute $x(2|2)$, the information is available to compute an updated predictor of $x(1)$ in the form of $x(1|2)$ that predicts the value of $x(1)$ using both $y(1)$, $y(2)$ rather than just $y(1)$ alone. Similarly, when $x(3|3)$ has been evaluated this necessarily means that the information is available to compute $x(1|3)$ and $x(2|3)$ that provide predictions of $x(1)$ and $x(2)$ using all the currently available responses $y(1)$, $y(2)$, $y(3)$. This particular process of updating predictors of previous state vectors for newly acquired responses is generally referred to as *smoothing*.

We will delve into the smoothing issue in more detail in the next chapter. However, the methods we will develop for smoothing there are backward recursions that are implemented after a forward pass through the data using Algorithm 4.3. Using the forward pass output $x(1|1)$, $S(1|1)$, $x(2|2)$, $S(2|2)$, ..., $x(n|n)$, $S(n|n)$, they further process it to obtain, e.g., $x(1|n)$, $S(1|n)$, $x(2|n)$, $S(2|n)$, ..., $x(n|n)$, $S(n|n)$. While this may be perfectly satisfactory in many cases, there are instances where more immediate updating may be useful.

In contrast to the algorithms developed in Chapter 5, the recursive method we will consider here provides "real time" updating of the signal and state vector predictors. That is, at any point in time t we will have a completely updated set of predictors $x(1|t)$, ..., $x(t|t)$. While our treatment of smoothing here is perhaps a bit premature, the algorithm we will develop is a forward recursion that can be carried out as an augmented version of the standard KF. Thus, it seems natural to discuss it at this stage of the overall presentation.

The computational scheme we will employ is really just an application of Algorithm 2.5 with a few additional details to transform the covariances computed there into the requisite vectors and matrices that are needed for constructing the state vector BLUPs and their prediction error variance-covariance matrices. To see the basic idea, let us begin with the second step in the forward KF Algorithm 4.3: i.e., after the completion of the $t = 2$ step in the **for** loop. At this point we have $x(1|1)$, $S(1|0)$, $S(1|1)$ that we obtained in the first step (i.e., after the initialization step of the algorithm) and $x(2|2)$, $S(2|1)$, $S(2|2)$. From $S(1|0)$ and $S(2|1)$ we can compute $M(1)$ and $M(2)$, respectively. Using Lemma 2.4 and the uncorrelated nature of the innovation vectors we then see that

$$x(1|2) = x(1|1)$$
$$+ S(1|0) M^T(1) H^T(2) R^{-1}(2) \varepsilon(2)$$
$$= x(1|1)$$
$$+ A(1) H^T(2) R^{-1}(2) \varepsilon(2)$$

for

$$A(1) = S(1|0) M^T(1)$$

and

$$S(1|2) = S(1|1)$$
$$- S(1|0) M^T(1) H^T(2) R^{-1}(2) H(2) M(1) S(1|0)$$
$$= S(1|1)$$
$$- A(1) H^T(2) R^{-1}(2) H(2) A^T(1).$$

At the next or $t = 3$ step the KF gives us $x(3|3)$, $S(3|2)$, $S(3|3)$ and we can use $S(3|2)$ to compute $M(3)$. Thus, let us now take $A(1)$ to be

$$A(1) = S(1|0) M^T(1) M^T(2),$$

which represents an "update" of our previous definition, and define the new matrix

$$A(2) = S(2|1) M^T(2).$$

Then,

$$x(1|3) = x(1|2)$$
$$+ A(1) H^T(3) R^{-1}(3) \varepsilon(3),$$
$$S(1|3) = S(1|2)$$
$$- A(1) H^T(3) R^{-1}(3) H(3) A^T(1),$$
$$x(2|3) = x(2|2)$$
$$+ A(2) H^T(3) R^{-1}(3) \varepsilon(3),$$
$$S(2|3) = S(2|2)$$
$$- A(2) H^T(3) R^{-1}(3) H(3) A^T(2).$$

On the $t = 4$ step the KF returns $x(4|4)$, $S(4|4)$, $M(4)$ and we update $A(1)$, $A(2)$ to

$$A(1) = S(1|0) M^T(1) M^T(2) M^T(3),$$

$$A(2) = S(2|1) M^T(2) M^T(3),$$

with the definition of the new matrix

$$A(3) = S(3|2) M^T(3).$$

Then, the updated predictors and prediction error variance-covariance matrices are

$$x(1|4) = x(1|3)$$
$$+ A(1) H^T(4) R^{-1}(4) \varepsilon(4),$$
$$S(1|4) = S(1|3)$$
$$- A(1) H^T(4) R^{-1}(4) H(4) A^T(1),$$
$$x(2|4) = x(2|3)$$
$$+ A(2) H^T(4) R^{-1}(4) \varepsilon(4),$$
$$S(2|3) = S(2|3)$$
$$- A(2) H^T(4) R^{-1}(4) H(4) A^T(2),$$
$$x(3|4) = x(3|3)$$
$$+ A(3) H^T(4) R^{-1}(4) \varepsilon(4),$$
$$S(3|4) = S(3|3)$$
$$- A(3) H^T(4) R^{-1}(4) H(4) A^T(3).$$

The special cases we have considered are enough to reveal patterns that we can exploit to carry out recursive computations. The algorithm below spells out some of the details for a particular implementation that can be developed from such considerations.

Algorithm 4.4 This algorithm computes $x(j|t)$, $S(j|t)$ for $t = 1, \ldots, n$ and $j = 1, \ldots, t$.

```
/*Initialization*/
```
$$S(1|0) = F(0) S(0|0) F^T(0) + Q(0)$$
$$R(1) = H(1) S(1|0) H^T(1) + W(1)$$
$$S(1|1) = S(1|0) - S(1|0) H^T(1) R^{-1}(1) H(1) S(1|0)$$
$$M(1) = F(1)$$
$$- F(1) S(1|0) H^T(1) R^{-1}(1) H(1)$$
$$\varepsilon(1) = y(1)$$

$$x(1|1) = S(1|0)H^T(1)R^{-1}(1)\varepsilon(1)$$
for $t = 2$ **to** n
$$S(t|t-1) = F(t-1)S(t-1|t-1)F^T(t-1)$$
$$+ Q(t-1)$$
$$R(t) = H(t)S(t|t-1)H^T(t) + W(t)$$
$$S(t|t) = S(t|t-1)$$
$$- S(t|t-1)H^T(t)R^{-1}(t)H(t)S(t|t-1)$$
$$M(t) = F(t)$$
$$- F(t)S(t|t-1)H^T(t)R^{-1}(t)H(t)$$
$$x(t|t-1) = F(t-1)x(t-1|t-1)$$
$$\varepsilon(t) = y(t) - H(t)x(t|t-1)$$
$$x(t|t) = S(t|t-1)H^T(t)R^{-1}(t)\varepsilon(t)$$
$$+ x(t|t-1)$$
/***Initialization of** $A(t-1)$*/
$$A(t-1) = S(t-1|t-2)$$
/*C **and** b **are temporary storage***/
$$C = H^T(t)R^{-1}(t)H(t)$$
$$b = H^T(t)R^{-1}(t)\varepsilon(t)$$
for $j = 1$ **to** $t - 1$
$$A(j) := A(j)M^T(t-1)$$
$$x(j|t) := x(j|t-1) + A(j)b$$
$$S(j|t) = S(j|t-1) - A(j)CA^T(j)$$
    end for
end for

    Algorithm 2.5 requires an overall computational effort of order n^2 flops. This is not surprising because the output of the algorithm consists of all the $n(n+1)/2$ state vector predictors that would be of interest from sequentially observed data: i.e., $x(1|1), \ldots, x(1|n), x(2|2), \ldots,$ $x(2|n), \ldots, x(n-1|n-1), x(n-1|n)$, and $x(n|n)$. In contrast, the smoothing algorithms in the next chapter are order n because they return only a subset of this predictor collection such as $x(1|n), x(2|n), \ldots, x(n|n)$.

4.5   Examples

In this section we will consider two examples that illustrate some of the work in the previous sections. First we
will look at the form of the KF recursions in the simple
state-space framework discussed in Sections 2.4 and 3.4.
Then we will implement Algorithm 4.3 to compute signal predictors for data with a Brownian motion signal as
discussed in Section 1.3.

Example: Time Invariant F, Q, H and W Matrices. Once again let
us restrict attention to the special case of a state-space
model with

$$y(t) = Hx(t) + e(t)$$

and

$$x(t + 1) = Fx(t) + u(t)$$

for known matrices H and F that do not depend on t
and e and u processes having time invariant variance-
covariance matrices Q_0 and W_0. Then, for example, (4.15)
and (4.20) simplify to

$$x(t|t - 1) = K(t - 1)\varepsilon(t - 1)$$

$$+ \sum_{j=1}^{t-2} F^{t-1-j} K(j)\varepsilon(j)$$

$$(4.21)$$

and

$$x(t|t) = S(t|t - 1)H^T R^{-1}(t)\varepsilon(t) + x(t|t - 1). \quad (4.22)$$

Let us focus on (4.21)–(4.22) in the case where $p = q = $
1 and use the work from Section 2.4 to analyze the form

of the coefficients or weights that are applied to the innovations when computing the predictor of $x(t)$. Now, in this context we can think of the $\varepsilon(j)$ as component variables that represent the information about the state $x(t)$ that is unique to the observation $y(j)$. This is intuitively clear from the way the innovations are constructed. But, we also have

$$x(t|t) = [x(t|t) - x(t|t-1)] + [x(t|t-1) - x(t|t-2)]$$
$$\cdots + [x(t|2) - x(t|1)] + x(t|1)$$

with

$$x(t|j) - x(t|j-1) = \mathrm{Cov}(x(t), \varepsilon(j))R^{-1}(j)\varepsilon(j).$$

This breaks the prediction of $x(t)$ into orthogonal pieces with $\varepsilon(j)$ providing the information about $x(t)$ that accrued from the realization of $y(j)$ beyond that which was already present at time index $j - 1$.

Now, using our large t approximations we have seen that

$$S(t|t-1) \approx \frac{R(\infty) - W_0}{H^2}$$

and

$$K(t) \approx \frac{F}{H}\left(1 - \frac{W_0}{R(\infty)}\right).$$

Applying these relations to (4.21)–(4.22) (while ignoring

the consequences of using them for small t) we obtain

$$x(t|t) \approx \frac{1}{H}\left(1 - \frac{W_0}{R(\infty)}\right)\varepsilon(t)$$

$$+ \frac{F}{H}\left(1 - \frac{W_0}{R(\infty)}\right)\varepsilon(t-1)$$

$$+ \sum_{j=1}^{t-2} F^{t-j}\frac{1}{H}\left(1 - \frac{W_0}{R(\infty)}\right)\varepsilon(j)$$

$$= \frac{1}{H}\left(1 - \frac{W_0}{R(\infty)}\right)\left[\varepsilon(t) + \sum_{j=1}^{t-1} F^{t-j}\varepsilon(j)\right].$$

$$(4.23)$$

In the case where $|F| < 1$, (4.23) suggests that when predicting $x(t)$ the information from a particular innovation is damped or down weighted as its time index becomes further below t. Put another way, in this case the innovations whose time indices are closest to t play more of a role in predicting $x(t)$.

Note that the factor F characterizes the memory properties of the state process since

$$x(t) = Fx(t-1) + u(t-1)$$

$$= F^2 x(t-2) + Fu(t-2) + u(t-1)$$

$$= F^3 x(t-3) + F^2 u(t-3)$$

$$+ Fu(t-2) + u(t-1),$$

etc., from which we can infer that when $|F| < 1$ the more recent states will have the most influence on the current state vector. From this perspective the KF's treatment of the innovations appears quite reasonable.

Example: Brownian Motion with White Noise. Now consider the case of sampling from Brownian motion with white noise that was introduced in Section 1.3. In this setting we sample at ordinates $0 \leq \tau_1 < \cdots < \tau_n \leq 1$, from the continuous time process

$$z(\tau) = B(\tau) + v(\tau), \quad \tau \in [0, 1],$$

where $v(\cdot)$ is a zero mean process with covariances

$$\text{Cov}(v(\tau), v(\nu)) = \begin{cases} W_0, & s = t, \\ 0, & s \neq t \end{cases}$$

and $B(\cdot)$ is a zero mean process with covariance kernel

$$\text{Cov}(B(\tau), B(\nu)) = \min(\tau, \nu).$$

We have already seen in Section 1.3 that this produces a state-space model with

$$y(t) = z(\tau_t)$$
$$= H(t)x(t) + e(t)$$

and

$$x(t+1) = B(\tau_{t+1})$$
$$= F(t)x(t) + u(t),$$

for

$$e(t) = v(\tau_t),$$
$$W(t) \equiv W_0,$$
$$H(t) \equiv 1,$$
$$u(t) = B(\tau_{t+1}) - B(\tau_t),$$
$$Q(t) = \tau_{t+1} - \tau_t,$$
$$F(t) \equiv 1$$

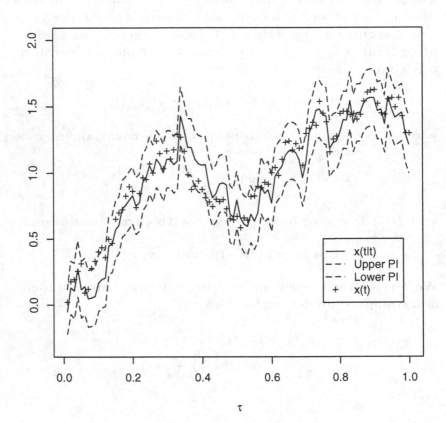

FIGURE 4.1

Predictions for Brownian motion with white noise data

with the convention that $\tau_0 := 0$.

Note that if we drop the restriction that the τ_t have to fall in the interval $[0, 1]$ and instead assume that $\tau_{t+1} - \tau_t = \Delta$ for some fixed constant Δ, then we are back in the framework of the previous example. In contrast to that case, here we are dealing with situations where the τ_t can be allowed to grow dense in $[0, 1]$.

Figure 4.1 shows the results returned from the KF in

the case of the sampled Brownian motion with white noise data from Figure 1.1. For ease of visual perception the values of $x(t|t)$ have been linearly connected as have the upper and lower bounds of the 95% prediction intervals (PI's)

$$x(t|t) \pm 2\sqrt{S(t|t)}.$$

The plus-signs in the plot are the true values of the signal: i.e., $f(t) = B(\tau_t)$.

5

Smoothing

5.1 Introduction

The KF Algorithm 4.3 produces the BLUP of $x(t)$ from $y(1), \ldots, y(t)$. Thus, at the end of the recursion we have calculated the vector of predictors

$$\begin{bmatrix} x(1|1) \\ x(2|2) \\ \vdots \\ x(n|n) \end{bmatrix}.$$

There are cases where this is exactly what is desired. For example, if only the current state vector at time t is of interest, then $x(t|t)$ gives the pertinent answer and the values of random vectors $x(1), \ldots, x(t-1)$ are no longer of relevance.

However, if previous values of the state vector are also objects of importance then it makes sense to include the information from new response data into their predictions as well. We discussed this briefly at the end of Chapter 4. Now we will develop the idea in more detail.

Because of the state-space structure, $y(t+1)$, $y(t+2)$, $\ldots, y(n)$ all contain information about the state vector $x(t)$. Consequently, by including new response information we can obtain better predictors of past states than would be obtained from using the filtering algorithm alone. This process of modifying the BLUP $x(t|t)$ based on $y(1), \ldots, y(t)$ to obtain a BLUP based on $y(1)$, $\ldots, y(t), y(t+1), \ldots, y(r)$ for some $r \geq t$ is referred to as *smoothing*.

The question then is how much smoothing to do or how much new information to incorporate into the estimator $x(t|t)$. One obvious answer would be to use the most information possible and predict the entire state vector via

$$
\begin{bmatrix} x(1|n) \\ x(2|n) \\ \vdots \\ x(n|n) \end{bmatrix}.
$$

This approach corresponds to a special case of what is called *fixed interval smoothing* that we will study in the next section.

5.2 Fixed interval smoothing

Suppose now that we have completed the forward pass through the data using the KF Algorithm 4.3. If that is the case, we now have at our disposal the quantities $x(t|t)$, $S(t|t)$ (as well as $S(t|t-1)$, $M(t)$, etc.) for $t = 1, \ldots, n$. The goal is to see how these predictors and prediction error variance-covariance matrices can be updated to incorporate information from responses that were not included in the forward pass or filtering run through the data.

The key result we will use to address smoothing questions is provided by the following.

Theorem 5.1 *For* $t < r \leq n$ *the BLUP of* $x(t)$ *based on* $y(1), \ldots, y(r)$ *is*

$$x(t|r) = x(t|t)$$

$$+ S(t|t-1) \sum_{j=t+1}^{r} A(t,j)\varepsilon(j)$$

$$(5.1)$$

with

$$A(t,j) = M^T(t) \cdots M^T(j-1) H^T(j) R^{-1}(j). \quad (5.2)$$

The prediction error variance-covariance matrix for $x(t|r)$ *is*

$$S(t|r) = S(t|t)$$

$$- S(t|t-1) \sum_{j=t+1}^{r} A(t,j) R(j) A^T(t,j) S(t|t-1).$$

$$(5.3)$$

The BLUP of $f(t)$ *based on* $y(1), \ldots, y(r)$ *is*

$$f(t|r) = H(t)x(t|r) \qquad (5.4)$$

with prediction error variance-covariance matrix

$$V(t|r) = H(t)S(t|r)H^T(t). \qquad (5.5)$$

Proof. The proof of this result is relatively straightforward given all our previous labor. First, observe from (1.29) that for $t < r \leq n$

$$x(t|r) = x(t|t)$$

$$+ \sum_{j=t+1}^{r} \text{Cov}(x(t), \varepsilon(j)) R^{-1}(j)\varepsilon(j).$$

Since the innovations are all uncorrelated, it follows from (1.30) that

$$S(t|r) = \text{Var}(x(t))$$

$$- \sum_{j=1}^{r} \text{Cov}(x(t), \varepsilon(j)) R^{-1}(j) \text{Cov}(\varepsilon(j), x(t))$$

$$= S(t|t)$$

$$- \sum_{j=t+1}^{r} \text{Cov}(x(t), \varepsilon(j)) R^{-1}(j) \text{Cov}(\varepsilon(j), x(t)).$$

An application of (2.18) from Lemma 2.4 now shows that

$$\text{Cov}(x(t), \varepsilon(j)) R^{-1}(j)$$

$$= S(t|t-1) M^T(t) \cdots M^T(j-1) H^T(j) R^{-1}(j)$$

$$= S(t|t-1) A(t, j)$$

which completes the argument.   ∎

Theorem 5.1 is stated in a way that is conducive to algorithmic development as a result of the "backdating" identity

$$A(t-1, j) = M^T(t-1) A(t, j). \tag{5.6}$$

Using (5.6) we see that

$$x(t-1|r) = x(t-1|t-1)$$

$$+ S(t-1|t-2) \sum_{j=t}^{r} A(t-1,j)\varepsilon(j)$$

$$= x(t-1|t-1) + S(t-1|t-2)A(t-1,t)\varepsilon(t)$$

$$+ S(t-1|t-2) \sum_{j=t+1}^{r} A(t-1,j)\varepsilon(j)$$

$$= x(t-1|t-1)$$

$$+ S(t-1|t-2)M^{T}(t-1)H^{T}(t)R^{-1}(t)\varepsilon(t)$$

$$+ S(t-1|t-2)M^{T}(t-1) \sum_{j=t+1}^{r} A(t,j)\varepsilon(j).$$

Consequently, we can accumulate the vector sums

$$\sum_{j=t+1}^{r} A(t,j)\varepsilon(j) \tag{5.7}$$

and use them to obtain a one step update of $x(t|t)$ to $x(t|r)$. A similar "backdating" result is true for the prediction error variance-covariance matrices since

$$A(t-1,j)R(j)A^{T}(t-1,j)$$

$$= M^{T}(t-1)A(t,j)R(j)A^{T}(t,j)M(t-1).$$

$$\tag{5.8}$$

The updating is then accomplished by accumulation of the matrix sums

$$\sum_{j=t+1}^{r} A(t,j)R(j)A^{T}(t,j).$$

The details involved in the recursive evaluation of $x(t|r)$,

$f(t|r)$, etc., are summarized in the algorithm below. Note that $x(r|r)$ is already returned from Algorithm 4.3.

Algorithm 5.1 This algorithm returns $x(t|r)$, $S(t|r)$, $f(t|r)$, $V(t|r)$, $t = 1, \ldots, r - 1$.

/*Initialization*/
$$a = M^T(r-1)H^T(r)R^{-1}(r)\varepsilon(r)$$
$$A = M^T(r-1)H^T(r)R^{-1}(r)H(r)M(r-1)$$
for $t = r - 1$ **to** 1
$$x(t|r) = x(t|t) + S(t|t-1)a$$
$$S(t|r) = S(t|t) - S(t|t-1)AS(t|t-1)$$
$$f(t|r) = H(t)x(t|r)$$
$$V(t|r) = H(t)S(t|r)H^T(t)$$
$$a = M^T(t-1)H^T(t)R^{-1}(t)\varepsilon(t) + M^T(t-1)a$$
$$A = M^T(t-1)H^T(t)R^{-1}(t)H(t)M(t-1)$$
$$\qquad + M^T(t-1)AM(t-1)$$
end for

Much like the Kalman filtering algorithm, the most salient feature of this recursion is its speed. The overall effort is $O(r)$ since the backdating steps require only one or two matrix multiplications and a vector or matrix addition. In particular, if we take $r = n$, the end result is $x(t|n)$, $S(t|n)$, $f(t|n)$, $V(t|n)$, $t = 1, \ldots, n - 1$, and the entire collection of these vectors and matrices can be evaluated in a total order n effort assuming that $x(t|t)$, $S(t|t)$, $M(t)$, $\varepsilon(t)$, $t = 1, \ldots, n - 1$, have already been computed. Since these latter quantities can also be obtained in order n operations using Algorithm 4.3, fixed interval smoothing with $r = n$ can be accomplished in $O(n)$ flops for state-space models.

Now, in general we know that for a nq-vector x and a np-vector y the BLUP of x based on y is provided by $\text{Cov}(x, y)\text{Var}^{-1}(y)y$. The computation of this predictor will require $O((np)^3)$ flops due to the need to "invert" the matrix $\text{Var}(y)$. Thus, the computation of the BLUP

for the state vector based on y can be accomplished two
orders of magnitude faster for state-space models than
would be the case for stochastic processes of a general
nature.

When $r = n$, Algorithm 5.1 is the same as the efficient smoothing algorithm in Kohn and Ansley (1989).
Prior to the development of this recursion fixed interval smoothing was accomplished using the relation (e.g.,
equation 4.5 on page 189 of Anderson and Moore 1979)

$$x(t|r) = x(t|t)$$
$$+ S(t|t-1)M^T(t)S^{-1}(t+1|t)[x(t+1|r)$$
$$- x(t+1|t)]. \qquad (5.9)$$

By initializing with $t = r - 1$, this identity can also be
used for fixed interval smoothing.

At first glance there is no reason to believe that (5.9)
is even true much less that by using it we would obtain
equivalent results to our other approach to fixed interval
smoothing. Perhaps of more importance, fixed interval s-
moothing methods derived from (5.9) appear to be slower
than Algorithm 5.1 (Kohn and Ansley 1989). Nonetheless, it is worthwhile to at least discuss this approach
and connect it to the developments in this section.

First note from (5.1)–(5.2) that (5.9) will be verified if
we can show that $M^T(t)S^{-1}(t+1|t)[x(t+1|r) - x(t+1|t)]$ is equal to the vector

$$\sum_{j=t+1}^{r} M^T(t) \cdots M^T(j-1)H^T(j)R^{-1}(j)\varepsilon(j).$$

But, from (5.1)–(5.2) and (4.20) we see that

$$x(t+1|r) - x(t+1|t)$$

$$= S(t+1|t) \sum_{j=t+2}^{r} A(t+1,j)\varepsilon(j)$$

$$+x(t+1|t+1) - x(t+1|t)$$

$$= S(t+1|t) \sum_{j=t+2}^{r} A(t+1,j)\varepsilon(j)$$

$$+S(t+1|t)H^T(t+1)R^{-1}(t+1)\varepsilon(t+1)$$

and (5.9) has been shown. At least one of the drawbacks
from using a recursion based on (5.9) becomes apparent
from this proof: namely, the inversion of $S(t+1|t)$ is un-
necessary in the sense that the desired quantities can be
computed directly using the accumulation of the vectors
(5.7) as in Algorithm 5.1.

Algorithm 5.1 is not the only way to approach the problem of signal prediction in the case of fixed interval s-
moothing with $r = n$. Instead, we could bypass the state
vector prediction step and use the previously derived i-
dentity

$$\hat{f} = y - W(L^T)^{-1}R^{-1}\varepsilon$$

to develop an algorithm for computing the BLUP for the
signal directly. Under this formulation efficient computation of the BLUP of f based on y is tantamount to
efficiently solving the system

$$L^T b = R^{-1}\varepsilon \tag{5.10}$$

for the vector b. To accomplish this one could, for example, use the structure in $(L^{-1})^T$ detailed in Theorem
3.2. Alternatively, one can merely finish the Cholesky solution of the linear system $\text{Var}(y)b = y$ by back-solving
the upper triangular system (5.10) with the help of The-

orem 3.1. We will briefly discuss both of these computational strategies.

As a result of Theorems 3.1 and 3.2 we have explicit expressions for the blocks of the upper triangular matrices L^T and $(L^T)^{-1}$. Using Theorem 3.1 we see that

$$b(n) = R^{-1}(n)\varepsilon(n),$$

$$b(n - 1) = R^{-1}(n - 1)\varepsilon(n - 1) - L^T(n, n - 1)b(n)$$

$$= R^{-1}(n - 1)\varepsilon(n - 1)$$

$$- K^T(n - 1)H^T(n)b(n),$$

$$b(n - 2) = R^{-1}(n - 2)\varepsilon(n - 2)$$

$$- L^T(n - 1, n - 2)b(n - 1)$$

$$- L^T(n, n - 2)b(n)$$

$$= R^{-1}(n - 1)\varepsilon(n - 1)$$

$$- K^T(n - 2)H^T(n - 1)b(n - 1)$$

$$- K^T(n - 2)F^T(n - 1)H^T(n)b(n)$$

and, in general,

$$b(t) = R^{-1}(t)\varepsilon(t) - K^T(t)[H^T(t + 1)b(t + 1)$$

$$+ \sum_{j=t+2}^{n} F^T(t + 1) \cdots F^T(j - 1)H^T(j)b(j)].$$

Similarly, using Theorem 3.2 we have

$$b(n) = R^{-1}(n)\varepsilon(n),$$

$$b(n-1) = R^{-1}(n-1)\varepsilon(n-1)$$
$$+(L^{-1})^T(n, n-1)R^{-1}(n)\varepsilon(n)$$
$$= R^{-1}(n-1)\varepsilon(n-1)$$
$$-K^T(n-1)H^T(n)R^{-1}(n)\varepsilon(n),$$

$$b(n-2) = R^{-1}(n-2)\varepsilon(n-2)$$
$$+(L^{-1})^T(n-1, n-2)R^{-1}(n-1)\varepsilon(n-1)$$
$$+(L^{-1})^T(n, n-2)R^{-1}(n)\varepsilon(n)$$
$$= R^{-1}(n-2)\varepsilon(n-2)$$
$$-K^T(n-2)H^T(n-1)R^{-1}(n-1)\varepsilon(n-1)$$
$$-K^T(n-2)M^T(n-1)H^T(n)R^{-1}(n)\varepsilon(n)$$

and, in general,

$$b(t) = R^{-1}(t)\varepsilon(t)$$
$$- K^T(t)[H^T(t+1)R^{-1}(t+1)\varepsilon(t+1)$$
$$+ \sum_{j=t+2}^{n} M^T(t+1)\cdots M^T(j-1)H^T(j)R^{-1}(j)\varepsilon(j)].$$

Consequently, the elements of $b = (L^T)^{-1}R^{-1}Ly$ can be computed efficiently in $O(n)$ operations by accumulating either of the vector sums

$$a_1(t) = H^T(t+1)b(t+1)$$
$$+ \sum_{j=t+2}^{n} F^T(t+1)\cdots F^T(j-1)H^T(j)b(j)$$

or

$$a_2(t) = H^T(t+1)R^{-1}(t+1)\varepsilon(t+1)$$

$$+ \sum_{j=t+2}^{n} M^T(t+1)\cdots M^T(j-1)H^T(j)R^{-1}(j)\varepsilon(j).$$

These sums are initialized with $a_1(n) = a_2(n) = 0$ and then updated using

$$a_1(t-1) = H^T(t)b(t) + F^T(t)a_1(t)$$

and

$$a_2(t-1) = H^T(t)R^{-1}(t)\varepsilon(t) + M^T(t)a_2(t).$$

Since the algorithms we have just described for computing the smoothed signal BLUP will produce exactly the same answer as the ones we would obtain from Algorithm 4.1 with $r = n$, we will not explore them in further detail here. Instead, we refer the reader to Eubank and Wang (2002) for a more complete development of the state-space model smoothing recursions from this perspective. The reason for introducing these alternative smoothing algorithms has been to shed light on the inner workings of the Kalman recursions. Specifically, we now see that, at least as far as signal estimation is concerned, the KF Algorithm 4.3 and smoothing Algorithm 5.1 when used in combination are equivalent to a smart Cholesky algorithm for solving $\text{Var}(y)b = y$. The resulting recursions use the structure of $\text{Var}(y)$ to return a solution two orders of magnitude faster than would be possible from a naive application of the Cholesky method. What makes this even more remarkable is that the linear system (5.10) is generally full with no banding or sparsity that can be exploited to lighten the computational burden.

There are other types of smoothing that can also be of interest in certain settings. For example, *fixed lag* smoothing concerns the computation of $x(t|t+k)$ for a fixed integer k. The idea in this case is that both t and n are growing and we wish to predict $x(t)$ using observations only up to k time index points into the "future." We will not discuss this or other smoothing concepts here apart from noting that Theorem 5.1 also provides a tool for developing recursive computational methods for these other smoothing scenarios. An overview of the various types of smoothing and associated KF algorithms can be found in de Jong (1989).

5.3 Examples

This section is essentially the smoothing parallel of Section 4.4. So, we will again examine the state-space model from Sections 2.4 and 3.4 and will also look at the effects of smoothing on the Brownian motion with white noise data from Section 1.3. For this latter example we will implement Algorithm 5.1 and compare the results to those obtained from Algorithm 4.3 thereby giving us a chance to visually examine the change in predictors that results from the updating process.

Example: Time Invariant F, Q, H and W matrices. Let us again examine the state-space model

$$y(t) = Hx(t) + e(t)$$

and

$$x(t + 1) = Fx(t) + u(t).$$

Due to the somewhat more complex formulas that arise in smoothing, it will be more useful to deal only with the situation where $p = q = 1$ so that H, F and the e and u process variances W_0, Q_0 are all scalars.

Using our large t approximations from Section 3.4 along with identity (5.1) the coefficient being applied to

$\varepsilon(j)$ in the smoothing step for prediction of $x(t)$ is

$$\frac{S(t|t-1)M(t)\cdots M(j-1)H}{R(j)}$$

$$\approx \frac{1}{H}\left[1 - \frac{W_0}{R(\infty)}\right]\left(\frac{FW_0}{R(\infty)}\right)^{j-t}.$$

If we combine this approximation with the one in (4.23) of Section 4.5 for $x(t|t)$ this produces, e.g.,

$$x(t|n) \approx \frac{1}{H}\left(1 - \frac{W_0}{R(\infty)}\right)\left[\varepsilon(t) + \sum_{j=1}^{t-1} F^{t-j}\varepsilon(j)\right]$$

$$+ \frac{1}{H}\left(1 - \frac{W_0}{R(\infty)}\right)\sum_{j=t+1}^{n}\left[\frac{FW_0}{R(\infty)}\right]^{j-t}\varepsilon(j).$$

Now $W_0/R(\infty) \leq 1$. So, in the "short memory" case for the state process where $|F| < 1$, we see that $x(t|n)$ represents a smoothing of data in the traditional sense of being (approximately) a weighted sum of the innovations with weights that are concentrated on innovations whose time indices are close to t.

Example: Brownian Motion with White Noise. Now consider again the case of sampling from Brownian motion with white noise that was introduced in Section 1.3. In this case we have $p = q = 1$, $H(t) = F(t) \equiv 1$, $W(t) \equiv W_0$ for some fixed constant W_0, $S(0|0) = 0$ and $Q(t) = \tau_t - \tau_{t-1}$ with $\tau_0 := 0$.

Data from this particular state space model were shown originally in Figure 1.1 that corresponded to the case where $W_0 = .025$. We applied the KF to that data in the second example of Section 4.5 with both the predictions and corresponding prediction intervals being shown in Figure 4.1. The parallel pointwise prediction results ob-

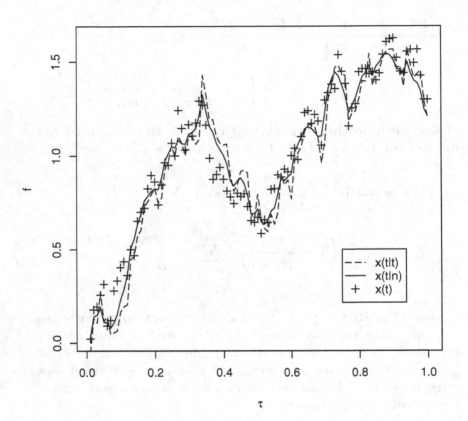

FIGURE 5.1

Predictor comparison for Brownian motion with white noise data

tained from fixed interval smoothing with $r = n = 100$ are compared to those from the KF in Figure 5.1. One can visually see the effect of smoothing from the plot and, as one would expect, the resulting "smoothed" predictions are better in that $\sum_{t=1}^{100} (x(t) - x(t|100))^2 = .714$ compared to $\sum_{t=1}^{100} (x(t) - x(t|t))^2 = 1.2877$.

Figure 5.2 shows the 95% prediction intervals for the

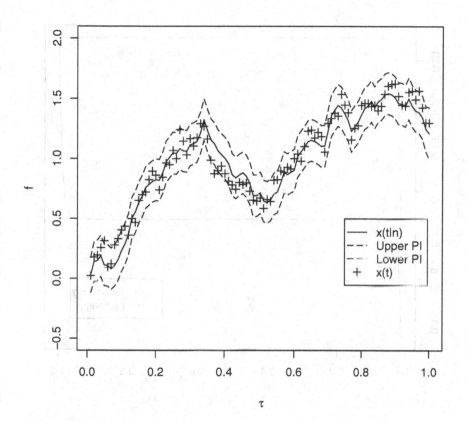

FIGURE 5.2

Smoothing Predictions for Brownian motion with white noise data

true signals $x(t)$ obtained via the formula

$$x(t|n) \pm 2\sqrt{S(t|n)}.$$

The coverage level for these particular prediction intervals was 96% compared to 92% for those obtained from the KF that are shown in Figure 4.1. More striking than

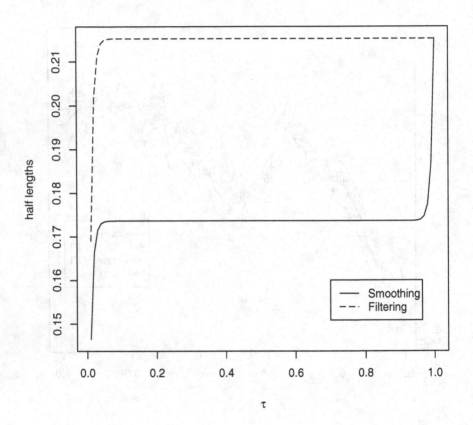

FIGURE 5.3

Prediction interval half lengths for filtering and smoothing

the difference in the coverage level is the difference in the width of the prediction intervals. This can be appreciated by examination of Figure 5.3 which shows the prediction interval half lengths for the two methods. The conclusion in this particular case is that the introduction of additional information in the backward smoothing recursion has substantially improved the precision of the predictions that were obtained in the initial filtering step.

6

Initialization

6.1 Introduction

In this chapter we discuss the initializing state vector $x(0)$ in more detail. We have assumed that $x(0)$ is a random vector with mean zero and variance-covariance matrix $S(0|0)$. As might be expected, the specific choice for $S(0|0)$ can have a profound effect on predictions since it affects all the subsequent variances and covariances as they are built up from the values of

$$\text{Var}(x(1)) = F(0)S(0|0)F^T(0) + Q(0),$$

$$\text{Cov}(x(1), \varepsilon(1)) = \text{Var}(x(1))H^T(1)$$

and

$$R(1) = \text{Var}(\varepsilon(1)) = H(1)\text{Var}(x(1))H^T(1) + W(1).$$

In the case where $\text{E}[x(0)] \neq 0$ there is also a location effect produced by the initial specification of the mean vector for $x(0)$. However, we will postpone treatment of this latter issue until Chapter 8.

There are some state-space models where there is a natural choice for $S(0|0)$. For instance, in the case of the Brownian motion example from Section 1.3 the $x(\cdot)$ process is staked at 0 which entails that $S(0|0) = 0$. However, in general, there may be no obvious choice for

$S(0|0)$ and, when that is the case, inference using state-space models would appear to be problematic. Ways to circumvent such difficulties include the use of diffuse (or improper prior) specifications for $x(0)$ and the treatment of $x(0)$ as an unknown, fixed quantity that must be estimated from the response data. It turns out that the two approaches lead to basically the same final answer. Thus, we will first deal with diffuse specifications and derivation of the so-called *diffuse Kalman filter*.

6.2   Diffuseness

One way to avoid the issue of specifying a distribution for $x(0)$ is to let its distribution be diffuse in a second order sense. That is, one takes $S(0|0)$ to be very (i.e., infinitely) large in a manner to be clarified shortly. The premise is that if we let the variances of the components of $x(0)$, i.e., the diagonal elements of $S(0|0)$, be large, then we have said virtually nothing about the sort of values $x(0)$ might actually take on.

To be somewhat more precise we will now take $x(0)$ to have mean 0 and variance-covariance matrix $S(0|0) = \nu I$. The corresponding (prior) probability distribution for $x(0)$ will then be allowed to become diffuse by letting $\nu \rightarrow \infty$. We will show that this produces predictions (and associated prediction error variance-covariances matrices) that do not involve $S(0|0)$ and, hence, can be computed directly without specifying anything about the variances or covariances for the initial state vector.

Obtaining freedom from the need for specification of initial conditions is not without its cost. One finds that our previous computational methodology can no longer be used to obtain predictions. Instead, it is necessary to modify the KF recursions to deal with changes in the predictors and their prediction error variance-covariance matrices that arise from using a diffuse prior for $x(0)$.

Although there are various ways to carry out the necessary algorithmic alterations, the resulting computational schemes are all generally referred to as *diffuse Kalman filters*.

To derive a diffuse KF let us first reformulate model (1.19)–(1.28) back into its signal-plus-noise parent model (1.1). To accomplish this we begin by observing that

$$x(1) = F(0)x(0) + u(0),$$

$$y(1) = H(1)F(0)x(0) + H(1)u(0) + e(1)$$

which progresses to

$$x(2) = F(1)x(1) + u(1)$$
$$= F(1)F(0)x(0) + F(1)u(0) + u(1),$$

$$y(2) = H(2)F(1)F(0)x(0) + H(2)F(1)u(0)$$
$$+ H(2)u(1) + e(2)$$

and then to

$$x(3) = F(2)x(2) + u(2)$$
$$= F(2)F(1)F(0)x(0) + F(2)F(1)u(0)$$
$$+ F(2)u(1) + u(2),$$

$$y(3) = H(3)F(2)F(1)F(0)x(0) + H(3)F(2)F(1)u(0)$$
$$+ H(3)F(2)u(1) + H(3)u(2) + e(3)$$

with the general case appearing as

$$
x(t) = F(t-1) \cdots F(0)x(0)
$$

$$
+ \sum_{j=0}^{t-2} F(t-1) \cdots F(j+1)u(j)
$$

$$
+ u(t-1),
$$

$$
y(t) = H(t)F(t-1) \cdots F(0)x(0)
$$

$$
+ \sum_{j=0}^{t-2} H(t)F(t-1) \cdots F(j+1)u(j)
$$

$$
+ H(t)u(t-1) + e(t).
$$

As a result of these calculations we can now express the response vector y and the state vector

$$
x = (x^T(1), \ldots, x^T(n))^T
$$

as

$$
y = HTx(0) + Hx_0 + e
$$

$$
= HTx(0) + f_0 + e \qquad (6.1)
$$

$$
x = Tx(0) + x_0 \qquad (6.2)
$$

with T the $nq \times q$ matrix

$$
T = \begin{bmatrix} F(0) \\ F(1)F(0) \\ \vdots \\ F(n-1) \cdots F(0) \end{bmatrix}, \qquad (6.3)
$$

x_0 the nq vector

$$x_0 =$$

$$\begin{bmatrix} u(0) \\ F(1)u(0) + u(1) \\ \vdots \\ \sum_{j=0}^{n-2} F(n-1) \cdots F(j+1)u(j) + u(n-1) \end{bmatrix},$$

$$(6.4)$$

H the $np \times nq$ matrix

$$H = \begin{bmatrix} H(1) & 0 & \cdots & 0 \\ 0 & H(2) & \cdots & 0 \\ \vdots & \vdots & \ddots & \vdots \\ 0 & 0 & \cdots & H(n) \end{bmatrix}, \qquad (6.5)$$

$f_0 = Hx_0$ and $e = (e^T(1), \ldots, e^T(n))^T$ defined as before.

Note that x_0 and f_0 are precisely the state and signal vectors we would have obtained from our state-space model if $x(0)$ could have been taken as zero in the sense of having $\mathrm{E}[x(0)] = 0$ and $S(0|0) = 0$. Among other things this means that there is a sub-model here where the ordinary KF Algorithms 4.3 and 5.1 for the case of $S(0|0) = 0$ could be applied directly. We need to figure out how this fact can be put to use in terms of allowing us to compute predictions for the model that actually generated the data.

The plan of attack is now as follows. First we will examine the form of predictors under the simple sub-model corresponding to x_0 in (6.4) that has $S(0|0) = 0$. Then, we will develop similar expressions for prediction in a more general case where we take $S(0|0) = \nu I$ for some $\nu > 0$. The next step in the process is to allow ν to grow large and thereby obtain a *diffuse* specification for $x(0)$.

In the limit as $\nu \to \infty$ we will have freed ourselves of the need to specify $S(0|0)$, provided that the limits exist for the BLUPs of x, f and their associated prediction error variance-covariance matrices. We will see that the limits are, in fact, well defined and, in addition, that they can be computed through an appropriate application of the ordinary fixed interval smoothing algorithm for the simple model having $S(0|0) = 0$. The resulting recursion provides one implementation of a diffuse KF.

6.2.1   Prediction when $S(0|0) = 0$

Let us now expand on our discussion of the state-space model with $S(0|0) = 0$. Thus, take

$$y_0 = Hx_0 + e_0$$

with x_0 structured along the lines of (6.4). To be precise, we are assuming that

1. e_0 is a random vector with components $e_0(t)$, $t = 1, \ldots, n$, having $\mathrm{E}[e_0] = 0$ and

$$\mathrm{Var}(e_0) = \mathrm{diag}(W(1), \ldots, W(n))$$

$$:= W, \qquad\qquad (6.6)$$

2.   the components of $y_0 = (y_0^T(1), \ldots, y_0^T(n))^T$ and $x_0 = (x_0^T(1), \ldots, x_0^T(n))^T$ follow the state-space model

$$y_0(t) = H(t)x_0(t) + e_0(t) \qquad (6.7)$$

$$x_0(t+1) = F(t)x_0(t) + u_0(t) \qquad (6.8)$$

with $u_0 = (u_0^T(0), \ldots, u_0^T(n-1))^T$ a zero mean random vector that is uncorrelated with

e_0 while having

$$\text{Var}(u_0) = \text{diag}(Q(0), \ldots, Q(n-1))$$
$$:= Q \tag{6.9}$$

and

3. the model is initialized by taking $\text{E}[x_0(0)] = 0$ and $S(0|0) = 0$.

Now we want to work out the form of the BLUPs of x_0 and $f_0 = Hx_0$ based on y_0. For this purpose we need the variance-covariance matrices for x_0 and y_0 as well as their cross-covariance matrix $\text{Cov}(x_0, y_0)$.

In view of (6.4) we can write

$$x_0 = Fu_0$$

with $F = \{F(t, j)\}_{t,j=1:n}$ an $nq \times nq$ lower triangular matrix having $F(t, t) = I$ and

$$F(t, j) = F(t-1) \cdots F(j) \tag{6.10}$$

for $j < t$. Thus,

$$\text{Var}(x_0) = FQF^T$$
$$:= \Sigma_{x_0} \tag{6.11}$$

and, hence,

$$\text{Var}(y_0) = H\Sigma_{x_0}H^T + W$$
$$:= \Sigma_{y_0}. \tag{6.12}$$

Finally, the x_0, y_0 cross-covariance matrix is

$$\text{Cov}(x_0, y_0) = \Sigma_{x_0}H^T. \tag{6.13}$$

Combining (6.11)–(6.13) with Theorem 1.1 we see that the least–squares predictors of x_0 and $f_0 = Hx_0$ based on y_0 are

$$\hat{x}_0(y_0) = \Sigma_{x_0} H^T (H\Sigma_{x_0} H^T + W)^{-1} y_0$$

$$= \Sigma_{x_0} H^T \Sigma_{y_0}^{-1} y_0 \qquad (6.14)$$

and

$$\hat{f}_0(y_0) = H\Sigma_{x_0} H^T (H\Sigma_{x_0} H^T + W)^{-1} y_0$$

$$= y_0 - W\Sigma_{y_0}^{-1} y_0$$

$$(6.15)$$

with the associated prediction error variance-covariance matrices for $\hat{x}_0(y_0)$ and $\hat{f}_0(y_0)$ given by

$$S_0 = \Sigma_{x_0} - \Sigma_{x_0} H^T \Sigma_{y_0}^{-1} H\Sigma_{x_0} \qquad (6.16)$$

and

$$V_0 = H\Sigma_{x_0} H^T$$

$$-H\Sigma_{x_0} H^T \Sigma_{y_0}^{-1} H\Sigma_{x_0} H^T$$

$$= W - W\Sigma_{y_0}^{-1} W, \qquad (6.17)$$

respectively. In deriving expressions concerning the signal BLUP we have used the fact that $H\Sigma_{x_0} H^T = \Sigma_{y_0} - W$. Thus, for example,

$$V_0 = (\Sigma_{y_0} - W) - (\Sigma_{y_0} - W)\Sigma_{y_0}^{-1} (\Sigma_{y_0} - W)$$

$$= (\Sigma_{y_0} - W)[I - \Sigma_{y_0}^{-1} (\Sigma_{y_0} - W)]$$

$$= W - W\Sigma_{y_0}^{-1} W.$$

The notation $\hat{x}_0(y_0)$ and $\hat{f}_0(y_0)$ undoubtedly seems a bit overly complicated. However, there is method to this notational madness in that we wish to think of relations

(6.14)–(6.15) as being prescriptions that tell us how to treat a particular input vector. Whether it makes sense to do so or not, we can certainly obtain vectors of "output" corresponding to any given $np \times 1$ "input" vector v through rote calculations using (6.14)–(6.15) with y_0 replaced by v. If v is not from our simple state-space model, then $\hat{x}_0(v)$ and $\hat{f}_0(v)$ would have no obvious interpretation. Nonetheless, they could be computed along with the diagonal blocks of S_0 and V_0 in order n operations using the fixed interval smoothing Algorithm 5.1 with $r = n$.

Of course, the efficient computation of $\hat{x}_0(y_0)$, $\hat{f}_0(y_0)$, etc., would seem to be of somewhat limited utility since it applies only to situations with $S(0|0) = 0$. As demonstrated by the examples in Section 1.3, cases where this is true are not uncommon. However, cases with $S(0|0) \neq 0$ can also arise as illustrated by the following example.

Example: Mean Shifted Brownian Motion with White Noise. Returning to our example of sampling from Brownian motion with white noise, the responses in that case had the form

$$y(t) = x(t) + e(t),$$

where $x(t) = B(\tau_t), t = 1, \ldots, n$, are random variables obtained from a Brownian motion process at the sampling points $0 \leq \tau_1 < \cdots < \tau_n \leq 1$ and $e(1), \ldots, e(n)$ are uncorrelated random variables with some common variance W_0 that are uncorrelated with $x(1), \ldots, x(n)$. For this model we necessarily have $x(0) = 0$ because the Brownian motion process vanishes with probability one at $\tau = 0$.

A generalization of the previous formulation would allow for $x(0) = \mu$ with μ a zero mean random variable having variance $S(0|0) = \nu$ that is uncorrelated with the Brownian motion signal and the random error terms $e(1), \ldots, e(n)$. The resulting state-space model would initialize with $x(1) = \mu + u(0)$ and $\mathrm{Var}(u(0)) = \tau_1$.

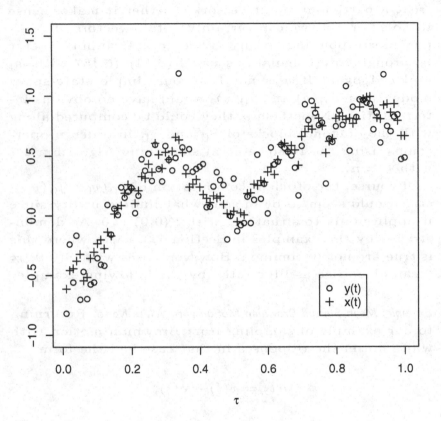

FIGURE 6.1

Mean shifted Brownian motion with white noise data

Then, in general, we would have

$$x(t+1) = x(t) + u(t)$$

$$= \mu + \sum_{j=0}^{t} u(j),$$

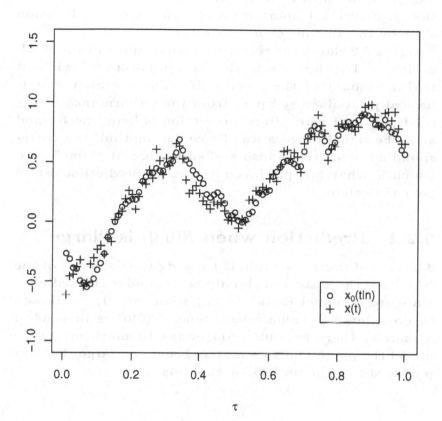

FIGURE 6.2

Naive predictions for mean shifted Brownian motion with white noise data

where the $u(t)$, $t = 0, \ldots, n-1$, are uncorrelated random variables having

$$\text{Var}(u(t)) = \tau_{t+1} - \tau_t$$

for $t = 1, \ldots, n-1$. This produces a version of Brownian motion that no longer has to be staked to 0 at "time" $\tau = 0$. An example of data obtained from this type of

model is shown in Figure 6.1. This is the same data we
saw in Figure 1.1 apart from the addition of a location
shift by the amount of $\mu = -.6364$.

Figure 6.2 shows the result of a naive application of Algorithm 5.1 with $r = n$ to the data in Figure 6.1 without
taking account of the mean shift. The overall result is
basically satisfactory apart from the performance of the
predictor near zero. Since prediction is being performed
as if the signal process was Brownian motion, the corresponding predictor is also staked to zero at "time" zero
which is what has produced the large prediction errors
near the origin.

6.2.2   Prediction when $S(0|0)$ is "large"

The goal of this subsection is to study the behavior of the
BLUPs of the state and signal vectors under the ordinary
state space model (1.19)–(1.28) when $S(0|0)$ is allowed
to grow large in some sense. Since $S(0|0)$ is in general
a matrix, there are numerous ways to mathematically
model the process of this matrix becoming large. For our
purposes there is no loss in choosing

$$S(0|0) = \nu I$$

and then letting $\nu \to \infty$. More general approaches such
as the one in de Jong (1991) actually lead to the same
formulae we will derive below.

The response and state vectors can now be written as
in (6.1)–(6.2) so that the BLUP of x based on y is

$$\hat{x}_\nu(y) = (\nu T G^T + \Sigma_{x_0} H^T)(\nu G G^T + \Sigma_{y_0})^{-1} y \quad (6.18)$$

with associated prediction error variance-covariance ma-

trix

$$S_\nu = \nu T T^T + \Sigma_{x0}$$

$$-(\nu T G^T + \Sigma_{x0} H^T)(\nu G G^T + \Sigma_{y0})^{-1}(\nu T G^T$$

$$+\Sigma_{x0} H^T)^T,$$

$$(6.19)$$

where

$$G = HT \qquad\qquad (6.20)$$

and $T, H, \Sigma_{x0}, \Sigma_{y0}$ are defined in (6.3), (6.5), (6.11) and (6.12). Concerning these quantities we are able to establish the following.

Theorem 6.1 *The BLUP of x when $S(0|0) = \nu I$ satisfies*

$$\lim_{\nu \to \infty} \hat{x}_\nu(y)$$

$$= [\Sigma_{x0} H^T \Sigma_{y0}^{-1}$$

$$-(\Sigma_{x0} H^T \Sigma_{y0}^{-1} G - T)(G^T \Sigma_{y0}^{-1} G)^{-1} G^T \Sigma_{y0}^{-1}]y$$

$$= \hat{x}_0(y)$$

$$-(\Sigma_{x0} H^T \Sigma_{y0}^{-1} G - T)(G^T \Sigma_{y0}^{-1} G)^{-1} G^T \Sigma_{y0}^{-1} y$$

$$:= \hat{x}_\infty \qquad\qquad (6.21)$$

and

$$\lim_{\nu \to \infty} S_\nu$$

$$= \Sigma_{x_0} - \Sigma_{x_0} H^T \Sigma_{y_0}^{-1} H \Sigma_{x_0}$$

$$+ [\Sigma_{x_0} H^T \Sigma_{y_0}^{-1} G$$

$$- T](G^T \Sigma_{y_0}^{-1} G)^{-1} [\Sigma_{x_0} H^T \Sigma_{y_0}^{-1} G - T]^T$$

$$= S_0 + [\Sigma_{x_0} H^T \Sigma_{y_0}^{-1} G$$

$$- T](G^T \Sigma_{y_0}^{-1} G)^{-1} [\Sigma_{x_0} H^T \Sigma_{y_0}^{-1} G - T]^T$$

$$:= S_\infty. \tag{6.22}$$

Let us first discuss the implications of this result before giving its proof. Formula (6.21) states that in the limit, as $\nu \to \infty$, the predictor $\hat{x}_\nu(y)$ can be obtained by a two step process. First one applies the formula (6.14) for prediction under the simple model having $S(0|0) = 0$ to the vector y that was obtained under the usual state-space model formulation (1.19)–(1.28) with $S(0|0) \neq 0$. This "predictor" is then corrected to work under the true model (i.e., when $S(0|0) \neq 0$ and "large") via subtraction of the term

$$(\Sigma_{x_0} H^T \Sigma_{y_0}^{-1} G - T)(G^T \Sigma_{y_0}^{-1} G)^{-1} G^T \Sigma_{y_0}^{-1} y.$$

The limiting prediction error variance-covariance matrix has a similar structure with a lead term $S_0 = \Sigma_{x_0} - \Sigma_{x_0} H^T \Sigma_{y_0}^{-1} H \Sigma_{x_0}$ from (6.16) corresponding to the simple model in Section 6.2.1 along with an adjustment term to make the ultimate answer conform to the actual diffuse specification of $S(0|0)$. The key here is that both $\hat{x}_0(y)$ and the diagonal blocks of S_0 can be obtained by a direct application of Algorithm 5.1 with $r = n$ to y by initializing with $S(0|0) = 0$. The corrections that are required for $\hat{x}_0(y)$ and S_0 can also be evaluated by fixed interval smoothing as we now demonstrate.

Suppose that we apply the standard forward KF Algorithm 4.3 having $S(0|0) = 0$ to $y(1), \ldots, y(n)$ and q additional "input variables" corresponding to the columns of the matrices

$$G(j) = H(j)F(j-1) \cdots F(0), \ j = 1, \ldots, n.$$

That is, at the tth step we carry out all of the steps to compute signal predictions, etc., for $q+1$ "response" vectors consisting of $y(t)$ and the q columns of the matrix $G(t)$. Then, at the end of the forward recursions we will have at our disposal R_0, the "innovation" vector for y,

$$\varepsilon_0 = L_0^{-1} y,$$

and another set of "innovation" vectors that comprise the columns of

$$E_0 = L_0^{-1} G, \qquad\qquad (6.23)$$

where L_0, R_0 are from the Cholesky decomposition

$$\Sigma_{y_0} = L_0 R_0 L_0^T.$$

This is because $\Sigma_{y_0} = \mathrm{Var}(y_0)$ is viewed as the response variance-covariance matrix when Algorithm 4.3 is applied with $S(0|0) = 0$ which entails that the forward KF recursion will return "innovation" vectors computed under this premise from any input vector including y and the columns of G. Thus, after the initial KF recursion we can efficiently compute (in order n operations) the matrix $G^T \Sigma_{y_0}^{-1} G = (GL_0^{-1})^T R_0^{-1} L_0^{-1} G = E_0^T R_0^{-1} E_0$ as well as the $q \times 1$ vector

$$
\begin{aligned}
g_0 &= G^T \Sigma_{y_0}^{-1} y \\
&= G^T (L_0^T)^{-1} R_0^{-1} L_0^{-1} y \\
&= E_0^T R_0^{-1} \varepsilon_0.
\end{aligned} \tag{6.24}
$$

One cautionary note is needed, however. To carry out the computations efficiently, the row blocks of G (i.e, the matrices $G(j) = H(j)F(j-1) \cdots F(0)$, $j = 1, \ldots, n$) must be evaluated recursively rather than by brute force. That is, we begin with $A(1) = F(0)$, $G(1) = H(1)A(1)$ and on the jth step for $j > 1$ we use the update $A(j) = F(j-1)A(j-1)$ to obtain $G(j) = H(j)A(j-1)$.

At the conclusion of the backward smoothing recursions (via Algorithm 5.1 with $S(0|0) = 0$ and $r = n$) for the response vector and the columns of G, we will have obtained $\hat{x}_0(y) = \Sigma_{x_0} H^T \Sigma_{y_0}^{-1} y$,

$$
\hat{G}_0 := \Sigma_{x_0} H^T \Sigma_{y_0}^{-1} G - T \tag{6.25}
$$

and the diagonal blocks of S_0 all in order n operations. Then, by solving the $q \times nq$ linear system

$$
\begin{aligned}
G^T \Sigma_{y_0}^{-1} G B &= G^T \Sigma_{y_0}^{-1} H \Sigma_{x_0} - T^T \\
&= \hat{G}_0^T \tag{6.26}
\end{aligned}
$$

for the $q \times nq$ matrix B one obtains the remainder of the pieces that are necessary to compute the diffuse estimators in (6.21): i.e., from this step we obtain

$$
\begin{aligned}
B &= (G^T \Sigma_{y_0}^{-1} G)^{-1} [G^T \Sigma_{y_0}^{-1} H \Sigma_{x_0} - T^T] \\
&= (E_0^T R_0^{-1} E_0)^{-1} \hat{G}_0^T. \tag{6.27}
\end{aligned}
$$

While there are many (i.e., nq) right-hand side columns in (6.26), the order of effort for each column is only q^3

so that the entire effort for obtaining B remains an order n calculation.

We can now express $\widehat{x}_\infty$ as

$$\widehat{x}_\infty = \widehat{x}_0(y) - B^T g_0 \qquad (6.28)$$

while the tth diagonal block of S_∞ is

$$S_\infty(t, t) = S_0(t, t) - \widehat{G}_0(t, 1:q) B(1:q, t), \qquad (6.29)$$

where $\widehat{G}_0(t, 1:q)$, $B(1:q, t)$ are, respectively, the tth row block and tth column block of $\widehat{G}_0$ in (6.25) and B in (6.27). Now all the calculations leading to expression (6.28) can be carried out in a total of order n flops. So, $\widehat{x}_\infty$ is also clearly returned in $O(n)$ flops and the same is true for the diagonal blocks of S_∞ or $S_\infty(t, t)$, $t = 1, \ldots, n$. This is because $S_0(t, t)$, $t = 1, \ldots, n$, are all returned from the fixed interval smoothing Algorithm 5.1 with $r = n$ in a total $O(n)$ effort and because evaluation of $\widehat{G}_0(t, 1:q) B(1:q, t)$ is only an order q^2 calculation for each t.

Proof of Theorem 6.1. The arguments for proving this theorem are algebraically tedious but conceptually straightforward. The first step involves an application of the the Sherman-Morrison-Woodbury formula (Householder 1964, pages 123–124) which states that

$$(A + BD^{-1}C)^{-1} = A^{-1}$$
$$-A^{-1}B(D - CA^{-1}B)^{-1}CA^{-1}$$

for matrices A, B, C and D. Applying this with $A = $

Σ_{y_0}, $B = G$, $C = G^T$ and $D^{-1} = \nu I$ gives

$$(\nu G G^T + \Sigma_{y_0})^{-1}$$

$$= \Sigma_{y_0}^{-1} - \Sigma_{y_0}^{-1} G [\nu^{-1} I + G^T \Sigma_{y_0}^{-1} G]^{-1} G^T \Sigma_{y_0}^{-1}$$

$$= \Sigma_{y_0}^{-1} - \Sigma_{y_0}^{-1} G (G^T \Sigma_{y_0}^{-1} G)^{-1} [I$$

$$+ (\nu G^T \Sigma_{y_0}^{-1} G)^{-1}]^{-1} G^T \Sigma_{y_0}^{-1}.$$

Next we use a matrix power series expansion for $(I + (\nu G^T \Sigma_{y_0}^{-1} G)^{-1})^{-1}$ to obtain

$$(\nu G G^T + \Sigma_{y_0})^{-1}$$

$$= \Sigma_{y_0}^{-1} - \Sigma_{y_0}^{-1} G (G^T \Sigma_{y_0}^{-1} G)^{-1} G^T \Sigma_{y_0}^{-1}$$

$$+ \nu^{-1} \Sigma_{y_0}^{-1} G (G^T \Sigma_{y_0}^{-1} G)^{-2} G^T \Sigma_{y_0}^{-1}$$

$$- \nu^{-2} \Sigma_{y_0}^{-1} G (G^T \Sigma_{y_0}^{-1} G)^{-3} G^T \Sigma_{y_0}^{-1} + O(\nu^{-3}).$$

The proof is concluded by direct multiplication of this expression by

$$(\nu T T^T + \Sigma_{x_0}) H^T = \nu T G^T + \Sigma_{x_0} H^T$$

and by

$$H(\nu T T^T + \Sigma_{x_0}) = \nu G T^T + H \Sigma_{x_0}.$$

Note that the columns of G are orthogonal to

$$\Sigma_{y_0}^{-1} - \Sigma_{y_0}^{-1} G (G^T \Sigma_{y_0}^{-1} G)^{-1} G^T \Sigma_{y_0}^{-1}$$

which directly eliminates certain terms of order ν and ν^2 from $\hat{x}_\nu(y)$ and S_ν. ∎

Since $\hat{f}_\nu(y) = H \hat{x}_\nu(y)$ and $V_\nu = \text{Var}(f - \hat{f}_\nu(y)) = H S_\nu H^T$ we also have an immediate corollary concerning the limiting properties of the signal predictor and its associated prediction error variance-covariance matrix.

Corollary 6.1 *Define $\hat{x}_\infty$ and S_∞ as in (6.21) and (6.22). Then, the BLUP of $f = Hx$ when $S(0|0) = \nu I$ satisfies*

$$\lim_{\nu \to \infty} \hat{f}_\nu(y) = H\hat{x}_\infty := \hat{f}_\infty \qquad (6.30)$$

and

$$\lim_{\nu \to \infty} V_\nu = HS_\infty H^T := V_\infty. \qquad (6.31)$$

We summarize the developments of this section by the following formulation of a diffuse Kalman filter algorithm.

Algorithm 6.1 This algorithm returns $\hat{x}_\infty$, $\hat{f}_\infty$ and $S_\infty(t, t)$, $V_\infty(t, t)$, $t = 1, \ldots, n$.

```
/*Inputs*/
```
$R_0 = \mathrm{diag}(R_0(0), \ldots, R_0(n))$
$E_0 = L_0^{-1} G$
$\varepsilon_0 = L_0^{-1} y$
$\hat{x}_0(y) = \Sigma_{x_0} H^T \Sigma_{y_0}^{-1} y$
$\hat{G}_0 = \Sigma_{x_0} H^T \Sigma_{y_0}^{-1} G - T$
$S_0(t, t), t = 1 : n$
```
/*Computation of x̂∞*/
```
$g_0 = E_0^T R_0^{-1} \varepsilon_0$
$A_0 = E_0^T R_0^{-1} E_0$
$B = A_0^{-1} \hat{G}_0^T$
$\hat{x}_\infty = \hat{x}_0(y) - (g_0^T B)^T$
```
/*Computation of f∞ and diagonal blocks
    of S∞, V∞*/
for t = 1 to n
```
    $S_\infty(t, t) = S_0(t, t) - \hat{G}_0(t, 1 : q)B(1 : q, t)$
    $\hat{f}_\infty(t) = H(t)\hat{x}_\infty(t)$
    $V_\infty(t, t) = H(t)S_\infty(t, t)H^T(t)$
```
end for
```

6.3   Diffuseness and least-squares estimation

There is an intimate connection between the predictors we have derived under a diffuse specification of $x(0)$ and those that would be obtained from least-squares "estimation" of $x(0)$. The two approaches really derive from fundamentally different assumptions about the initial state vector. But, they produce the same answer which means that the diffuse KF can be used to efficiently compute predictions in the case where $x(0)$ is "estimated" and, conversely, we could have discovered the form of the diffuse predictors directly from an "estimation" perspective.

Let us now be more precise about the "estimation" viewpoint. The idea in this case is that $x(0)$ is a fixed, unknown vector of parameters. If we disallow degenerate random vectors whose probability distributions are concentrated on a single point in $\mathcal{R}^q$, then this is completely different from the treatment of $x(0)$ in previous sections.

The model now has the form

$$y = Hx + e$$

$$= G\beta + HFu + e, \tag{6.32}$$

$$x = T\beta + Fu \tag{6.33}$$

with

$$u = (u^T(0), \ldots, u^T(n-1))^T,$$

e as before and H and F as defined in (6.5) and (6.10). Note that we have replaced the $x(0)$ notation by β in (6.32)–(6.33) to signify that we are now viewing it as a fixed, albeit unknown, parameter rather than an element of the $x(\cdot)$ process.

Under model (6.32)–(6.33) the response vector follows a (generalized) linear model in that we can write

$$y = G\beta + \tilde{e}$$

with the random "error" vector $\tilde{e} = HFu + e$ having a variance-covariance matrix that is not proportional to the identity. The Gauss-Markov Theorem tells us how to deal with such situations and provides us with the *best linear unbiased estimator* (BLUE) of β: namely,

$$(G^T \text{Var}^{-1}(\tilde{e})G)^{-1} G^T \text{Var}^{-1}(\tilde{e})y.$$

But, $\text{Var}(\tilde{e})$ is just our old friend Σ_{y0} and, hence, the BLUE of β is

$$\hat{\beta} = (G^T \Sigma_{y0}^{-1} G)^{-1} G^T \Sigma_{y0}^{-1} y. \qquad (6.34)$$

This expression certainly looks familiar and we can readily detect its presence in the formulae for $\hat{x}_\infty$ and $\hat{f}_\infty$ that appear in Theorem 6.1 and Corollary 6.1.

To take this one step further, note that if β were known, then Theorem 1.1 would have the consequence that the BLUP of x would be

$$T\beta + \Sigma_{x0} H^T \Sigma_{y0}^{-1} (y - G\beta).$$

Since β is unknown one way to deal with the situation would be to use this BLUP formula with β replaced by the BLUE in (6.34). This results in the predictor

$$T\hat{\beta} + \Sigma_{x0} H^T \Sigma_{y0}^{-1} (y - G\hat{\beta})$$

$$= T(G^T \Sigma_{y0}^{-1} G)^{-1} G^T \Sigma_{y0}^{-1} y$$

$$+ \Sigma_{x0} H^T \Sigma_{y0}^{-1} [I - G(G^T \Sigma_{y0}^{-1} G)^{-1} G^T \Sigma_{y0}^{-1}] y$$

which is now recognized as being identical to the diffuse BLUP $\hat{x}_\infty$ of x provided by (6.21) in Theorem 6.1.

Unlike the development that led to our derivation of $\widehat{x}_\infty$ from the standard state-space model setting, our re-discovery of $\widehat{x}_\infty$ as a predictor for model (6.32)–(6.33) is basically *ad hoc*. While we know that $\widehat{x}_\infty$ is at least the limit of optimal (i.e., BLUP) predictors under state-space model (1.19)–(1.28), no such qualities can be immediately attributed to the predictor when it is used under model (6.32)–(6.33). The remainder of this section will be devoted to filling in this void. Specifically we will show that x_∞ is a BLUP predictor of x under model (6.32)–(6.33) as well, in the sense that it minimizes

$$\mathrm{E}(x - Ay)^T (x - Ay) \qquad (6.35)$$

over all predictors of the form Ay that satisfy the unbiasedness condition

$$\mathrm{E}[Ay] = T\beta \qquad (6.36)$$

for all $\beta \in \mathcal{R}^q$.

Our main result about $\widehat{x}_\infty$ is given in Theorem 6.2 below. Results on BLUPs in this type of mixed model setting are far from new. A classical reference on the subject is provided by Goldberger (1962) while more modern treatments can be found in the discussion article by Robinson (1991) and references therein.

Theorem 6.2 *The BLUP of x in model (6.33) based on y in model (6.32) is $\widehat{x}_\infty$ in (6.21). Its prediction error variance-covariance matrix under model (6.32)–(6.33) is S_∞ in (6.22).*

Proof. To begin let us write $\widehat{x}_\infty$ in its linear estimator form with $\widehat{x}_\infty = A_\infty y$ for

$$A_\infty = \Sigma_{x_0} H^T \Sigma_{y_0}^{-1}$$

$$+ T(G^T \Sigma_{y_0}^{-1} G)^{-1} G^T \Sigma_{y_0}^{-1}$$

$$- \Sigma_{x_0} H^T \Sigma_{y_0}^{-1} G (G^T \Sigma_{y_0}^{-1} G)^{-1} G^T \Sigma_{y_0}^{-1}.$$

Note that $A_\infty G = T$ which has the implication that $\hat{x}_\infty$ is unbiased as a predictor of x. More generally, the condition that (6.36) must hold for all $\beta \in \mathcal{R}^q$ entails that we must have $AG = T$ for any unbiased linear predictor of the form $\hat{x} = Ay$.

Now we must show that the choice of $A = A_\infty$ minimizes (6.35). For this purpose, note that as in the proof of Theorem 1.1 we have the identity

$$E(x - Ay)^T (x - Ay) = E(x - A_\infty y)^T (x - A_\infty y)$$
$$+ 2E(x - A_\infty y)^T (A_\infty y - Ay)$$
$$E(A_\infty y - Ay)^T (A_\infty y - Ay).$$

Thus, the first part of the Theorem will be proved if we can demonstrate that

$$E(x - A_\infty y)^T (A_\infty - A)y = 0.$$

Since both A_∞ and A satisfy (6.36) it follows that $(A_\infty - A)y = (A_\infty - A)(HFu + e)$. Consequently,

$$E(x - A_\infty y)^T (A_\infty - A)y$$
$$= Ex^T (A_\infty - A)(HFu + e)$$
$$- E(HFu + e)^T A_\infty^T (A_\infty - A)(HFu + e)$$
$$= \mathrm{tr}[H\Sigma_{x_0} (A_\infty - A)]$$
$$- \mathrm{tr}[\Sigma_{y_0} A_\infty^T (A_\infty - A)]$$
$$= \mathrm{tr}[(H\Sigma_{x_0} - \Sigma_{y_0} A_\infty^T)(A_\infty - A)]$$

with tr again representing the trace functional for matrices. In obtaining the last expression we used the cyclic property of the trace and the facts that $\mathrm{Cov}(HFu + e, x) = H\Sigma_{x_0}$ and $\Sigma_{y_0} = \mathrm{Var}(HFu + e)$.

A direct calculation gives

$$H\Sigma_{x_0} - \Sigma_{y_0} A_\infty^T$$

$$= G(G^T \Sigma_{y_0}^{-1} G)^{-1} G^T \Sigma_{y_0}^{-1} H\Sigma_{x_0}$$

$$-G(G^T \Sigma_{y_0}^{-1} G)^{-1} T.$$

Thus, from the cyclic property of the trace

$$\mathrm{tr}[(H\Sigma_{x_0} - \Sigma_{y_0} A_\infty^T)(A_\infty - A)]$$

$$= \mathrm{tr}[(A_\infty - A)(H\Sigma_{x_0} - \Sigma_{y_0} A_\infty^T)]$$

$$= 0$$

because $(A_\infty - A)G = 0$.

It remains to show that

$$\mathrm{E}(x - \hat{x}_\infty)(x - \hat{x}_\infty)^T = S_\infty$$

under model (6.32)–(6.33). This result follows after some relatively straightforward algebraic manipulations once one realizes that

$$x - \hat{x}_\infty = Fu - A_\infty[HFu + e].$$

■

There is nothing special about estimation of x and similar results to those in Theorem 6.1 hold for estimation of f. We state this formally in the following corollary to conclude the section.

Corollary 6.2 *The BLUP of* $f = Hx$ *for model (6.32)–(6.33) is* $\hat{f}_\infty$ *in (6.30). Its prediction error variance-covariance matrix under model (6.32)–(6.33) is* V_∞ *in (6.31).*

6.4 An example

Let us now return to the mean shifted Brownian motion example from the end of Section 6.2.1 for which the responses are shown along with the true signals (or, equivalently, states in this case) in Figure 6.1. The data were generated from a model of the form

$$y(t) = x(t) + e(t),$$

where the $e(t)$ are uncorrelated random errors with common variance W_0 and the state process is determined by

$$x(1) = \mu + u(0),$$

and

$$x(t + 1) = \mu + x_0(t) + u(t)$$

with $x_0(t) = \sum_{j=0}^{t-1} u(j)$ for uncorrelated random variables $u(t), t = 0, \ldots, n - 1$, having common variance Q_0. The data in Figure 6.1 correspond to the choices $W_0 = .025$, $Q_0 = 1/100$ and $\mu = -.6364$.

For this example $p = q = 1$ and $H(t) = F(t - 1) = 1, t = 1, \ldots, n$. This means that $H = I$ in (6.5) while F in (6.10) is a lower triangular matrix of all unit elements. The matrices (6.11) and (6.12) are then found to be

$$\Sigma_{x_0} = F Q F^T = \{Q_0 \min(i, j)\}_{i,j=1:n}$$

and

$$\Sigma_{y_0} = \{Q_0 \min(i, j) + W_0\}_{i,j=1:n}.$$

In this instance $G = T = 1$, where 1 is a $n \times 1$ vector of all unit elements.

Using formula (6.34) from Section 6.3 with μ now playing the role of β we obtain an "estimator" of the mean

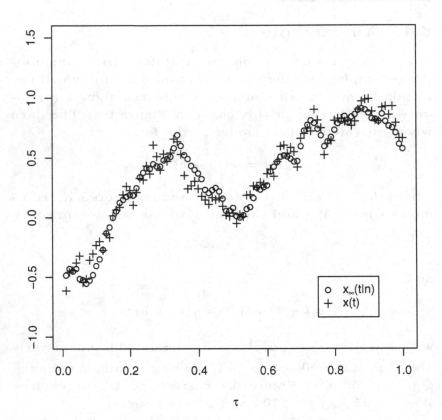

FIGURE 6.3

Predictions for mean shifted Brownian motion with white noise data

shift as $\widehat{\mu} = 1^T \Sigma_{y_0}^{-1} y / 1^T \Sigma_{y_0}^{-1} 1$. Using this result in formula (6.21) then gives

$$\widehat{x}_\infty = \widehat{x}_0(y) - \widehat{\mu}(\Sigma_{x_0} \Sigma_{y_0}^{-1} 1 - 1).$$

For the data in Figure 6.1 we find that $\widehat{\mu} = -.5766$. The resulting vector of predictions $\widehat{x}_\infty$ is shown in Fig-

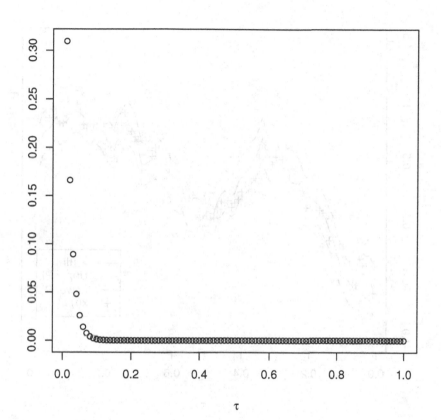

FIGURE 6.4

Correction to naive estimator

ure 6.3 along with the true shifted Brownian motion signal. By comparing with the naive estimator $\widehat{x}_0(y)$ in Figure 6.2 we see that primary differences between the two
sets of predictions occurs in the area around $\tau = 0$ where
the nonmean corrected estimator attempts to mimic the
Brownian motion signal and vanish at the origin. Figure
6.4 plots the correction factor $\widehat{\mu}(\Sigma_{x_0} \Sigma_{y_0}^{-1} 1 - 1)$ that is
subtracted from $\widehat{x}_0(y)$. From this we can see that the

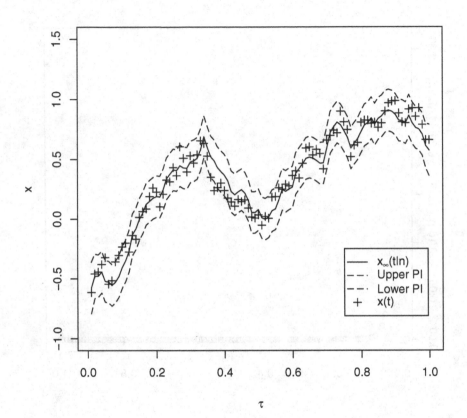

FIGURE 6.5

Prediction intervals with x_∞

adjustments being made to $\hat{x}_0$ have a local nature in
that the correction term essentially vanishes at sampling
points $\tau_j = j/n$ that are very far removed from zero.

Finally, Figure 6.5 shows the 95% prediction interval-
s obtained using x_∞ in conjunction with the diagonal
elements of S_∞ in (6.22). The actual coverage level in
this case was 96%. A comparison of the half lengths of
the S_∞ based intervals with those one would have ob-

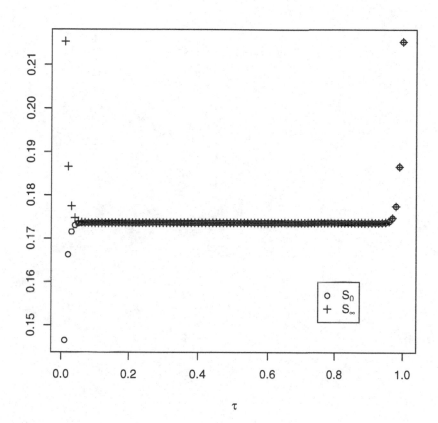

FIGURE 6.6

Prediction interval half lengths using S_0 and S_∞

tained by incorrectly using intervals obtained from S_0 is provided by Figure 6.6. Again, the effect of the mean correction is local with larger intervals arising from the mean corrected approach only around the lower boundary region where $\tau = 0$.

7

Normal Priors

7.1 Introduction

State-space models clearly have a Bayesian connection since one can view $x(t), t = 0, \ldots, n$, as parameters whose prior distributions are determined from the state equation (1.20). When the $u(\cdot)$ and $x(0)$ vectors are specified to be normal, as is often the case in Bayesian settings, the responses are also normal provided that the $e(\cdot)$ are normally distributed. In any case, whether one views state-space models from a Bayesian or frequentist perspective, there will be a closed form expression for the sample likelihood when both the $x(\cdot)$ and $e(\cdot)$ processes are normal.

The KF Algorithm 4.3 has some additional utility for normal state-space models. In that instance it can be used to efficiently evaluate the likelihood function which, in turn, can be employed for estimation of any unknown parameters. In this chapter we explore how the KF can be used for this purpose.

7.2   Likelihood evaluation

We now want to obtain an expression for the sample likelihood under a normal state-space model. To accomplish this we will proceed as in Chapter 6 and rewrite the response and state equations in vector-matrix form: i.e., we have

$$y = Gx(0) + HFu + e,$$

$$x = Tx(0) + Fu,$$

where

$$T = \begin{bmatrix} F(0) \\ F(1)F(0) \\ \vdots \\ F(n-1)\cdots F(0) \end{bmatrix},$$

$$H = \begin{bmatrix} H(1) & 0 & \cdots & 0 \\ 0 & H(2) & \cdots & 0 \\ \vdots & \vdots & \ddots & \vdots \\ 0 & 0 & \cdots & H(n) \end{bmatrix},$$

$$G = HT,$$

F is a block lower triangular matrix with identity diagonal blocks and below diagonal blocks given by

$$F(t, j) = F(t-1)\cdots F(j), \ j < t,$$

$e = (e^T(1), \ldots, e^T(n))^T$ and $u = (u^T(0), \ldots, u^T(n-1))^T$.

The zero mean, random vectors u, e and $x(0)$ are now all taken to be mutually independent and normally distributed with $\text{Var}(x(0)) := S(0|0)$,

$$\text{Var}(u) = \text{diag}(Q(0), \ldots, Q(n-1))$$

$$:= Q,$$

$$\text{Var}(e) = \text{diag}(W(1), \ldots, W(n))$$

$$:= W.$$

Consequently, the state vector is normally distributed with mean zero and variance-covariance matrix

$$\text{Var}(x) = TS(0|0)T^T + FQF^T$$

and, similarly, the response vector y has a normal probability distribution with zero mean and variance-covariance matrix

$$\text{Var}(y) = H\text{Var}(x)H^T + W.$$

The density for y is therefore seen to be

$$f(y) = \frac{1}{(2\pi)^{pn/2}|\text{Var}(y)|^{1/2}} \exp\left\{-\frac{y^T \text{Var}^{-1}(y)y}{2}\right\}.$$

$$(7.1)$$

As a result of (7.1), two times the negative of the logarithm of the sample likelihood is seen to be

$$\ell = pn \ln 2\pi + \ln |\text{Var}(y)| + y^T \text{Var}^{-1}(y)y.$$

We will refer to this quantity subsequently as simply the *log-likelihood*. Our aim at this point is to devise a computationally efficient method for evaluating the log-likelihood.

If we again write the y variance-covariance matrix in its Cholesky factorized form

$$\text{Var}(y) = LRL^T,$$

then we know that the vector of innovations is $\varepsilon = L^{-1}y$. Also, $|\text{Var}(y)| = |L||R||L^T| = |R|$ since both L^T and L are triangular with identity matrix diagonal blocks. Thus,

$$\ell - pn \ln 2\pi = \ln|R| + \varepsilon^T R^{-1}\varepsilon$$

$$= \sum_{t=1}^{n} \ln|R(t)|$$

$$+ \sum_{t=1}^{n} \varepsilon^T(t)R^{-1}(t)\varepsilon(t) \qquad (7.2)$$

and we conclude that the entire likelihood can be evaluated directly in order n operations at the end of the forward pass of the KF. In fact, we need not actually employ the full-blown KF Algorithm 4.3 since all the information we need to compute ℓ is returned in Algorithm 4.1, for example.

It also follows that by using Algorithm 4.1 we will be able to adaptively update the log-likelihood function in cases where response vectors are actually being observed as a time ordered sequence. More precisely, if

$$\ell_t - tp \ln 2\pi = \sum_{k=1}^{t} [\ln|R(k)| + \varepsilon^T(k)R^{-1}(k)\varepsilon(k)]$$

is the log-likelihood after seeing observations $y(1), \ldots, y(t)$ then

$$\ell_{t+1} = \ell_t + p \ln 2\pi$$

$$+ \ln|R(t+1)| + \varepsilon^T(t+1)R^{-1}(t+1)\varepsilon(t+1)$$

is the log-likelihood for $y(1), \ldots, y(t+1)$.

The following modified version of Algorithms 2.1 and 4.1 evaluates the sample likelihood in order n flops.

Algorithm 7.1 This algorithm evaluates the likelihood less a factor of $pn \ln 2\pi$.

/*Initialization*/
$$S(1|0) = F(0)S(0|0)F^T(0) + Q(0)$$
$$R(1) = H(1)S(1|0)H^T(1) + W(1)$$
$$S(1|1) = S(1|0) - S(1|0)H^T(1)R^{-1}(1)H(1)S(1|0)$$
$$K(1) = F(1)S(1|0)H^T(1)R^{-1}(1)$$
$$\varepsilon(1) = y(1)$$
$$A(1) = K(1)\varepsilon(1)$$
$$\ell = \ln|R(1)| + \varepsilon(1)^T R^{-1}(1)\varepsilon(1)$$
for $t = 2$ **to** $n - 1$
$$S(t|t-1) = F(t-1)S(t-1|t-1)F^T(t-1)$$
$$+Q(t-1)$$
$$R(t) = H(t)S(t|t-1)H^T(t) + W(t)$$
$$S(t|t) = S(t|t-1)$$
$$-S(t|t-1)H^T(t)R^{-1}(t)H(t)S(t|t-1)$$
$$K(t) = F(t)S(t|t-1)H^T(t)R^{-1}(t)$$
$$\varepsilon(t) = y(t) - H(t)A(t-1)$$
$$A(t) = K(t)\varepsilon(t) + F(t)A(t-1)$$
$$\ell = \ell + \ln|R(t)| + \varepsilon(t)^T R^{-1}(t)\varepsilon(t)$$
end for
$$S(n|n-1) = F(n-1)S(n-1|n-1)F^T(n-1)$$
$$+Q(n-1)$$
$$R(n) = H(n)S(n|n-1)H^T(n) + W(n)$$
$$\varepsilon(n) = y(n) - H(n)A(n-1)$$
$$\ell = \ell + \ln|R(n)| + \varepsilon(n)^T R^{-1}(n)\varepsilon(n)$$

All of the developments in this section assume that $S(0|0)$ has been specified. We can relax this assumption by allowing for diffuse specifications for $x(0)$ as in the next section.

7.3   Diffuseness

Let us now adapt the results from the previous section to deal with situations where the distribution of $x(0)$ is unknown and allowed to become diffuse. To accomplish this we will proceed along the same lines as we did in Section 6.2.2.

When $S(0|0) = \nu I$ we have

$$\mathrm{Var}(y) = \nu H T T^T H^T + H F Q F^T H^T + W$$

$$= \nu G G^T + \Sigma_{y0}$$

with

$$\Sigma_{y0} := H F Q F^T H^T + W.$$

From arguments in Section 6.3 we know that

$$(\nu G G^T + \Sigma_{y0})^{-1} = \Sigma_{y0}^{-1}$$

$$- \Sigma_{y0}^{-1} G (G^T \Sigma_{y0}^{-1} G)^{-1} G^T \Sigma_{y0}^{-1}$$

$$+ O(\nu^{-1}).$$

Consequently, we can apply the KF innovation Algorithm 4.1 for $S(0|0) = 0$ to $y(1)$, ..., $y(n)$ to obtain the vector of "innovations"

$$\varepsilon_0 = L_0^{-1} y$$

with L_0 deriving from the Cholesky decomposition

$$\Sigma_{y0} = L_0 R_0 L_0^T$$

and $R_0 = (R_0(1), \ldots, R_0(n))$ the Cholesky matrix factors that represent the variance-covariance matrices for

$\varepsilon_0(1), \ldots, \varepsilon_0(n)$ when $\nu = 0$. Then,

$$\lim_{\nu \to \infty} y^T \operatorname{Var}(y)^{-1} y$$

$$= y^T (\Sigma_{y0}^{-1}$$

$$- \Sigma_{y0}^{-1} G (G^T \Sigma_{y0}^{-1} G)^{-1} G^T \Sigma_{y0}^{-1}) y$$

$$= \varepsilon_0^T (R_0^{-1}$$

$$- R_0^{-1} E_0 (E_0^T R_0^{-1} E_0)^{-1} E_0^T R_0^{-1}) \varepsilon_0$$

with

$$E_0 = L_0^{-1} G.$$

This shows that the innovation part of the likelihood in the diffuse case can be computed in order n operations by applying Algorithm 4.1 to both y and the q columns of the $np \times q$ matrix $G = HT$.

It remains to consider $|\nu G G^T + \Sigma_{y0}|$. Again, using $|L_0| = |L_0^T| = 1$ we have

$$|\nu G G^T + \Sigma_{y0}| = |R_0||\nu E_0 E_0^T R_0^{-1} + I|$$

$$= |R_0||\nu E_0^T R_0^{-1} E_0 + I|$$

using the identity $|I + CD| = |I + DC|$ which is valid provided the products CD and DC are defined. But,

$$|\nu E_0^T R_0^{-1} E_0 + I|$$

$$= |\nu I||E_0^T R_0^{-1} E_0 + \nu^{-1} I|$$

$$= |\nu I||E_0^T R_0^{-1} E_0||I + \nu^{-1} (E_0^T R_0^{-1} E_0)^{-1}|$$

$$= \nu^q |E_0^T R_0^{-1} E_0|(1 + o(1))$$

as $\nu \to \infty$.

Upon combining all our approximations we see that

$$\lim_{\nu \to \infty} [\ell - pn \ln 2\pi - q \ln \nu]$$

$$= \ln |E_0^T R_0^{-1} E_0|$$

$$+ \sum_{t=1}^{n} \left[\ln |R_0(t)| + \varepsilon_0^T(t) R_0^{-1}(t) \varepsilon_0(t) \right]$$

$$- \varepsilon_0^T R_0^{-1} E_0 (E_0^T R_0^{-1} E_0)^{-1} E_0^T R_0^{-1} \varepsilon_0$$

$$:= \ell_\infty. \tag{7.3}$$

This entire quantity can be computed in order n operations using a modified version of Algorithm 7.1. To be a bit more specific, the diffuse log-likelihood can be evaluated efficiently as follow:

1. Obtain $\varepsilon_0 = L_0^{-1} y$, $E_0 = L_0^{-1} G$, R_0 by applying Algorithm 4.1 with $S(0|0) = 0$ to both y and the columns of G.

2. Follow the computational scheme in Algorithm 7.1 (also with $S(0|0) = 0$) to calculate

$$\ell_0 = \sum_{t=1}^{n} [\ln |R_0(t)| + \varepsilon_0^T(t) R_0^{-1}(t) \varepsilon_0(t)].$$

3. Compute the vector

$$g_0 = E_0^T R_0^{-1} \varepsilon_0,$$

solve the linear system

$$(E_0^T R_0^{-1} E_0) b_0 = E_0^T R_0^{-1} \varepsilon_0$$

for the $q \times 1$ vector b_0 and evaluate the determinant

$$d_0 = |G^T \Sigma_{y_0}^{-1} G| = |E_0^T R_0^{-1} E_0|.$$

4.   The diffuse likelihood is then obtained from

$$\ell_\infty = \ell_0 + \ln d_0 - b_0^T g_0.$$

One additional conclusion that can be drawn from this is that smoothing considerations do not arise in the context of likelihood evaluation even in the diffuse case.

There is a bit of subterfuge involved in using (7.3) that arises from our removal of the term $q \ln \nu = \ln |\nu I|$ from the log-likelihood before taking the limit. Certainly the limit would not be defined if we did not make such an adjustment. However, there is a question as to whether this tactic can be placed on a more solid theoretical foundation. The remainder of this section is devoted to providing some rigorous justification for this approach.

Let J be a $np \times (np - q)$ matrix satisfying $J^T G = 0$ and $J^T J = I$. Then, following Ansley and Kohn (1985), consider the transformation from y to a new vector of variables

$$\omega = \begin{bmatrix} \omega_1 \\ \omega_2 \end{bmatrix} = \begin{bmatrix} J^T y \\ G^T y \end{bmatrix}.$$

This transformed vector variate is normal with mean zero and variance-covariance matrix

$$\begin{bmatrix} J^T \Sigma_{y_0} J & J^T \Sigma_{y_0} G \\ G^T \Sigma_{y_0} J & \nu B^2 + G^T \Sigma_{y_0} G \end{bmatrix}$$

with $B = G^T G$. From this we see that the conditional density of ω_1 given ω_2, $f(\omega_1 | \omega_2)$, is that of a normal random vector with mean

$$J^T \Sigma_{y_0} G (\nu B^2 + G^T \Sigma_{y_0} G)^{-1} \omega_2$$

$$= \nu^{-1} J^T \Sigma_{y_0} G B^{-1} (I$$

$$+ \nu^{-1} B^{-1} G^T \Sigma_{y_0} G B^{-1}) B^{-1} \omega_2$$

and variance-covariance matrix

$$J^T \Sigma_{y_0} J - \nu^{-1} J^T \Sigma_{y_0} G B^{-1} (I$$

$$+ \nu^{-1} B^{-1} G^T \Sigma_{y_0} G B^{-1})^{-1} B^{-1} G^T \Sigma_{y_0} J.$$

Since $(I + \nu^{-1} B^{-1} G^T \Sigma_{y_0} G B^{-1})^{-1} = I + O(\nu^{-1})$ we can conclude that

$$\lim_{\nu \to \infty} f(\omega_1 | \omega_2) = f(\omega_1), \tag{7.4}$$

where $f(\omega_1)$ is the unconditional density of the vector ω_1 which has a normal distribution with mean 0 and variance-covariance matrix $J^T \Sigma_{y_0} J$.

Similarly, by observing that $\nu B^2 + G^T \Sigma_{y_0} G = \nu B^2 + O(1)$ we have that

$$\nu^{q/2} f(\omega_2)$$

$$= \frac{\nu^{q/2} \exp \left\{ -\dfrac{\omega_2^T (\nu B^2 + G^T \Sigma_{y_0} G)^{-1} \omega_2}{2} \right\}}{(2\pi)^{q/2} |\nu B^2 + G^T \Sigma_{y_0} G|^{1/2}}$$

$$\to \frac{1}{(2\pi)^{q/2} |B|^2}$$

as $\nu \to \infty$. Combining all these results reveals that the y density $f(y) = f(\omega_1 | \omega_2) f(\omega_2)$ satisfies

$$\lim_{\nu \to \infty} \nu^{q/2} f(y) = \frac{f(\omega_1)}{(2\pi)^{q/2} |B|^2}. \tag{7.5}$$

Note that this result is invariant with respect to J meaning that the specific choice of transformation is immaterial provided only that $J^T G = 0$.

Tracing back through all our arguments we see that we began by transforming to two new vector random variables: namely, $\omega_1 = J^T y = J^T [HFu + e]$ which is independent of $x(0)$ and a second random vector $\omega_2 = G^T y$

whose covariance with $x(0)$ is $\nu G^T G$. Consequently, we partitioned out everything from the responses that depended on $x(0)$ and placed it in ω_2 while the part of y that contained no information about $x(0)$ was segregated into ω_1. By using *only* the new random variable ω_1 (and ignoring ω_2) we can conduct inference about the process without having to concern ourselves with the initial state vector. Although discarding the information from ω_2 will entail some loss of estimation efficiency, this particular strategy is by no means unprecedented. For example the same basic tactic shows up in the context of variance components estimation where $Gx(0)$ plays the role of a fixed effect. See, e.g., Hocking (2003).

Our final result was the observation that in the limit the sample likelihood when scaled by a factor of $\nu^{-q/2}$ is proportional to the unconditional density of $f(\omega_1)$. But, this is equivalent to saying that

$$\lim_{\nu \to \infty}[-\ln(f(y)) - \frac{q}{2}\ln\nu] = -\ln(f(\omega_1)) + c$$

with c a factor that does not involve ω_2. Consequently, if we use the the diffuse likelihood for inference about parameters, etc., this will provide us with exactly the same answer that we would have obtained by working with the transformed variable ω_1 that is independent of the initial state vector.

7.4   Parameter estimation

In general a state-space model may involve unknown parameters that appear in transition matrices, variance-covariance matrices or arise from other sources. For normal state-space models these parameters can be estimated by maximization of the likelihood function. In the case of a specified variance-covariance matrix for the initial state vector one would use the parameter values that minimize

$$\sum_{t=1}^{n} [\ln |R(t)| + \varepsilon^{T}(t)R^{-1}(t)\varepsilon(t)].$$

For the diffuse setting we would similarly minimize

$$\sum_{t=1}^{n} [\ln |R_0(t)| + \varepsilon_0^{T}(t)R_0^{-1}(t)\varepsilon_0(t)]$$

$$+ \ln \left| E_0^{T} R_0^{-1} E_0 \right|$$

$$- \varepsilon_0^{T} R_0^{-1} E_0 (E_0^{T} R_0^{-1} E_0)^{-1} E_0^{T} R_0^{-1} \varepsilon_0.$$

There is one situation of interest where minimization of the sample log-likelihood function can be carried out explicitly: namely, when there is a common scale parameter σ^2 that appears in the e, u and $x(0)$ variances and covariances. In such a case we will have

$$\mathrm{Var}(x(0)) = \sigma^2 S(0|0),$$

$$\mathrm{Var}(u) = \sigma^2 Q,$$

$$\mathrm{Var}(e) = \sigma^2 W,$$

so that

$$\mathrm{Var}(y) = \sigma^2 \left(GS(0|0)G^{T} + HFQF^{T}H^{T} + W \right)$$

$$= \sigma^2 \left(GS(0|0)G^{T} + \Sigma_{y0} \right)$$

and

$$\ell - pn \ln 2\pi = pn \ln \sigma^2 + \ln \left| GS(0|0)G^{T} + \Sigma_{y0} \right|$$

$$+ \frac{y^{T} \left(GS(0|0)G^{T} + \Sigma_{y0} \right)^{-1} y}{\sigma^2}.$$

It follows from this that the *maximum likelihood estimator* (mle) of σ^2 is

$$\hat{\sigma}^2 = \frac{y^T \left(GS(0|0)G^T + \Sigma_{y0} \right)^{-1} y}{np}.$$

By replacing σ^2 with $\hat{\sigma}^2$ in the log-likelihood we obtain (apart from an additive constant) the *concentrated* log-likelihood function

$$\ln |GS(0|0)G^T + \Sigma_{y0}| + pn \ln \hat{\sigma}^2$$

that can be further optimized to effect estimation of other parameters. A similar analysis for the diffuse setting produces

$$\hat{\sigma}^2 = \frac{y^T \left(\Sigma_{y0}^{-1} - \Sigma_{y0}^{-1} G(G^T \Sigma_{y0}^{-1} G)^{-1} G^T \Sigma_{y0}^{-1} \right) y}{np}$$

(7.6)

as the mle of σ^2 with

$$\ln \left| G^T \Sigma_{y0}^{-1} G \right| + \ln \left| \Sigma_{y0} \right| + pn \ln \hat{\sigma}^2 \qquad (7.7)$$

representing the concentrated log-likelihood.

As a final development let us return to the setting of Section 6.3 where $x(0)$ was taken to be a vector of fixed parameters rather than a random vector. In that case we had

$$y = G\beta + HFu + e$$

with β a $q \times 1$ vector of unknown parameters that occupies the place of the initializing state vector in this formulation.

When u and e are normal, y is normal with mean $G\beta$ and variance-covariance matrix Σ_{y0}. Thus, two times the negative log-likelihood is (apart from additive constants)

$$\ln |\Sigma_{y0}| + (y - G\beta)^T \Sigma_{y0}^{-1} (y - G\beta).$$

From this we obtain the mle of β as

$$\widehat{\beta} = (G^T \Sigma_{y0}^{-1} G)^{-1} G^T \Sigma_{y0}^{-1} y$$

which is identical to the BLUE of β obtained in Section 6.3. By replacing β with $\widehat{\beta}$ in the log-likelihood function we arrive at the "concentrated" log-likelihood (sans constants)

$$\ln \left| \Sigma_{y0} \right| + (y^T - G\widehat{\beta})^T \Sigma_{y0}^{-1} (y - G\widehat{\beta})$$

$$= \ln \left| \Sigma_{y0} \right| + y^T \Sigma_{y0}^{-1} y$$

$$- y^T \Sigma_{y0}^{-1} G (G^T \Sigma_{y0}^{-1} G)^{-1} G^T \Sigma_{y0}^{-1} y$$

which agrees with the diffuse log-likelihood apart from the absence of the factor $\ln |G^T \Sigma_{y0}^{-1} G|$. One conclusion to be drawn from this is that Algorithm 4.1 can also be employed to efficiently evaluate the concentrated log-likelihood that arises when the initial state vector is viewed as a vector of parameters that is estimated via the method of maximum likelihood.

7.5   An example

Let us return to our mean shifted Brownian motion example from Section 6.4 and consider the problem of likelihood based parameter estimation in that context. The response and state vector in this instance are

$$y = x + e, \tag{7.8}$$

$$x = \mu 1 + F u \tag{7.9}$$

with F a $n \times n$ lower triangular matrix of all unit elements and 1 a $n \times 1$ vector of all unit elements. We will now assume that the u and e vectors are composed of independent normal random variables with variances Q_0 and σ^2 (rather than W_0 as we have used previously) and define

$$\Sigma_{y0} = \{\lambda_0 \min(i, j) + 1\}_{i,j=1:n}$$

with $\lambda_0 = Q_0/\sigma^2$.

The mle of σ^2 for data from (7.8)–(7.9) is now seen to be

$$\hat{\sigma}^2 = y^T \Sigma_{y0}^{-1} y - \frac{(1^T \Sigma_{y0}^{-1} y)^2}{1^T \Sigma_{y0}^{-1} 1}.$$

The concentrated (with respect to σ^2) log-likelihood is then a function of λ alone that may be minimized to obtain an estimator for the value of λ_0. This can be accomplished as follows. Define

$$\Sigma(\lambda) = \{\lambda \min(i, j) + 1\}_{i,j=1:n}$$

and

$$\sigma^2(\lambda) = y^T \Sigma^{-1}(\lambda) y - \frac{(1^T \Sigma^{-1}(\lambda) y)^2}{1^T \Sigma^{-1}(\lambda) 1}.$$

Then, our estimator of λ_0 is provided by the value $\hat{\lambda}$ that minimizes, with respect to λ, the function

$$\ell(\lambda) = \ln 1^T \Sigma^{-1}(\lambda) 1 + \ln |\Sigma(\lambda)| + n \ln \sigma^2(\lambda)$$

assuming that $\hat{\lambda}$ is unique.

Applying the above results to the data in Figure 6.1 produces the estimator $\hat{\sigma}^2 = .02618$ which is to be compared to the actual parameter value of $\sigma^2 = .025 = W_0$ in this case. For this data we also had $Q_0 = .01$ so that

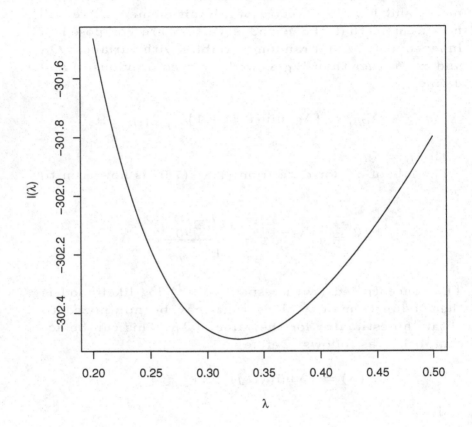

FIGURE 7.1

Concentrated log likelihood for mean shifted Brownian motion data

$\lambda_0 = .4$. Figure 7.1 shows a plot of the concentrated log-likelihood $\ell(\lambda)$ as a function λ that was evaluated using the KF innovation recursion applied to y and the vector 1 (along the lines of discussions in Section 7.3) for each value of λ in a grid of points distributed over $[.2, .5]$. From the plot we can see that the mle of λ_0 is $\hat{\lambda} \approx .33$.

8

A General State-Space Model

8.1   Introduction

Our discussions up to now have focused on the model
having $p \times 1$ response vectors

$$y(t) = H(t)x(t) + e(t)$$

corresponding to $q \times 1$ state vectors

$$x(t + 1) = F(t)x(t) + u(t),$$

where $H(t)$, $F(t)$, $t = 1, \ldots, n$, are known matrices, $u(0)$,
$\ldots, u(n - 1)$ are zero mean, random q-vectors that are
uncorrelated with each other and with the uncorrelated,
zero mean, random p-vector $e(1), \ldots, e(n)$. The covariance structure of the model is then determined from the
specifications

$$\text{Var}\,(u(t)) = Q(t),\ t = 0, \ldots, n - 1,$$

$$\text{Var}\,(e(t)) = W(t),\ t = 1, \ldots, n,$$

for known matrices $Q(t-1)$, $W(t)$, $t = 1, \ldots, n$, and the
condition that the initial state vector has zero mean and

$$\text{Var}\,(x(0)) = S(0|0)$$

for a known matrix $S(0|0)$. To conclude our treatment of the KF we want to expand this model somewhat to allow for more general applications.

Perhaps the first thing to realize is that our previous restriction that the $y(\cdot)$ and $x(\cdot)$ vectors all have fixed dimensions p and q, respectively, has never been necessary. Indeed, all the recursions we have developed up to now will still work perfectly well with dimensions that change with the t index. Thus, from this point on we can proceed as if $y(t)$ and $x(t)$ are, respectively, $p_t \times 1$ and $q_t \times 1$ with $H(t), F(t), W(t), Q(t)$ now representing $p_t \times q_t$, $q_{(t+1)} \times q_t$, $p_t \times p_t$ and $q_t \times q_t$ matrices, respectively.

The next step is to broaden our original model formulation by allowing for nonzero means. To accomplish this we will use a response equation of the form

$$y(t) = A_Y(t)\beta + H(t)x(t) + e(t) \qquad (8.1)$$

for $t = 1, \ldots, n$, coupled with the state equation

$$x(t+1) = A_X(t)\beta + F(t)x(t) + u(t), \qquad (8.2)$$

for $t = 0, \ldots, n-1$, where $A_Y(t), A_X(t-1), t = 1, \ldots, n$, are, respectively, $p_t \times r$ and $q_t \times r$ known matrices, β is a r-vector of parameters, the $e(\cdot)$ and $u(\cdot)$ processes are as before and $x(0) = 0$ in the sense that

$$x(1) = A_X(0)\beta + u(0). \qquad (8.3)$$

As a result of (8.3) the mean vector for $x(1)$ is $\mathrm{E}[x(1)] = A_X(0)\beta$ which has the consequence that

$$\mathrm{E}[y(1)] = A_Y(1)\beta + H(1)A_X(0)\beta.$$

Similarly, the mean for $x(2)$ is $\mathrm{E}[x(2)] = A_X(1)\beta + F(1)A_X(0)\beta$ so that

$$\mathrm{E}[y(2)] = A_Y(2)\beta + H(2)A_X(1)\beta + H(2)F(1)A_X(0)\beta.$$

In general, the mean for $x(t)$ is

$$E[x(t)] = A_X(t-1)\beta$$

$$+ \sum_{j=1}^{t-1} F(t-1) \cdots F(j) A_X(j-1)\beta$$

$$(8.4)$$

and the mean for $y(t)$ is

$$E[y(t)] = A_Y(t)\beta + H(t)A_X(t-1)\beta$$

$$+ H(t) \sum_{j=1}^{t-1} F(t-1) \cdots F(j) A_X(j-1)\beta$$

$$= A_Y(t)\beta + H(t)E[x(t)]. \qquad (8.5)$$

Model (8.1)–(8.3) reverts back to (1.19)–(1.28) with $S(0|0) = 0$ when $\beta = 0$. To recover (1.19)–(1.28) with $S(0|0) \neq 0$ we would need to allow for β to be random. This approach will generally lead to questions concerning parameters in the prior distribution for β that ultimately lead to a diffuse specification as in Section 6.2.2. Such an approach can be carried out for the present setting as well with de Jong (1991) providing a thorough treatment of this type of development. Here we will begin by assuming that β is fixed and known, initially. Then, in the next section, we derive its BLUE and thereby obtain the BLUPs of state and signal vectors for the case that β is unknown. As we showed in Section 7.4 for a somewhat simpler model, it follows from de Jong (1988, 1991) that the resulting predictors and their prediction error variance-covariance matrices are identical to the ones obtained via use of a diffuse prior.

Another common state-space model has response equations of the form

$$y(t) = A_Y(t)\beta + H(t)x(t) + G_Y(t)v(t)$$

and state equations

$$x(t+1) = A_X(t)\beta + F(t)x(t) + G_X(t)z(t),$$

where $v(\cdot), z(\cdot)$ are zero mean processes with identity variance-covariance matrices. Situations such as this can be converted to our original formulation involving the $e(\cdot)$ and $u(\cdot)$ processes by taking $W(t) = G_Y(t)G_Y^T(t)$ and $Q(t) = G_X(t)G_X^T(t)$. Consequently, we will forego the notational overhead of treating this case and instead focus our attention on (8.1)–(8.3).

8.2   KF recursions

The first step in extending our work from Chapters 4–5 to model (8.1)–(8.3) is to define an innovation process in this more general setting. To accomplish this we will proceed much as we did before except for the use of a mean correction. That is, we begin with

$$\varepsilon(1) = y(1) - E[y(1)], \tag{8.6}$$

$$R(1) = \text{Var}(\varepsilon(1)) \tag{8.7}$$

and then take

$$\varepsilon(t) = y(t) - E[y(t)]$$
$$- \sum_{j=1}^{t-1} \text{Cov}(y(t), \varepsilon(j)) R^{-1}(j)\varepsilon(j), \tag{8.8}$$

$$R(t) = \text{Var}(\varepsilon(t)), \tag{8.9}$$

for $t = 2, \ldots, n$. Note that it is necessarily the case that $E[\varepsilon(t)] = 0, t = 1, \ldots, n$.

Now that we have the innovations, we can obtain the BLUPs of $x(t)$ and the signal vector $f(t) = H(t)x(t)$

through the use of Theorem 1.1. Specifically, because the innovations in (8.6) and (8.8) have zero means and are uncorrelated, we obtain

$$x(t|j) = E[x(t)]$$

$$+ \sum_{i=1}^{j} \text{Cov}(x(t), \varepsilon(i)) R^{-1}(i) \varepsilon(i) \quad (8.10)$$

$$f(t|j) = E[f(t)]$$

$$+ \sum_{i=1}^{j} \text{Cov}(f(t), \varepsilon(i)) R^{-1}(i) \varepsilon(i)$$

$$= H(t) x(t|j). \quad (8.11)$$

Because variances and covariances are mean invariant, it is necessarily the case that the variances and covariances for the innovations as well as their covariance with the state vectors under model (8.1)–(8.3) can be computed *exactly* as before under model (1.19)–(1.28) for the case where $S(0|0) = 0$. In particular, this means that

$$S(1|0) = \text{Var}(x(1)) = Q(0),$$

$$R(1) = \text{Var}(\varepsilon(1)) = H^{T}(1) Q(0) H(1) + W(1),$$

$$S(1|1) = Q(0) - Q(0) H^{T}(1) R^{-1}(1) H(1) Q(0)$$

and, for $t > 1$,

$$S(t|t) = S(t|t-1)$$

$$-S(t|t-1)H^T(t)R^{-1}(t)H(t)S(t|t-1),$$

$$S(t|t-1) = F(t-1)S(t-1|t-1)F^T(t-1)$$

$$+Q(t-1),$$

$$R(t) = H(t)S(t|t-1)H^T(t) + W(t),$$

$$\text{Cov}(x(t), \varepsilon(t)) = S(t|t-1)H^T(t).$$

Also, from Lemma 2.4 we will have

$$\text{Cov}(x(t), \varepsilon(j)) = F(t-1)\cdots F(j)S(j|j-1)H^T(j),$$

for $j \leq t-1$ and

$$\text{Cov}(x(t), \varepsilon(j))$$

$$= S(t|t-1)M^T(t)M^T(t+1)\cdots M^T(j-1)H^T(j)$$

for $j \geq t+1$.

Let us delve a bit further into the issue of why our previous covariance formulae remain valid under model (8.1)–(8.3). We may write

$$x(1) = A_X(0)\beta + u(0),$$

$$y(1) = A_Y(1)\beta + H(1)A_X(0)\beta + H(1)u(0) + e(1),$$

$$x(2) = A_X(1)\beta + F(1)A_X(0)\beta + F(1)u(0) + u(1),$$

$$y(2) = A_Y(2)\beta + H(2)A_X(1)\beta$$

$$+H(2)F(1)A_X(0)\beta + H(2)F(1)u(0) + H(2)u(1)$$

$$+e(2)$$

and, in general,

$$x(t) = A_X(t-1)\beta \tag{8.12}$$

$$+ \sum_{j=1}^{t-1} F(t-1)\cdots F(j)A_X(j-1)\beta$$

$$+ \sum_{j=1}^{t-1} F(t-1)\cdots F(j)u(j-1) + u(t-1)$$

$$= \mathrm{E}[x(t)] + \sum_{j=1}^{t-1} F(t-1)\cdots F(j)u(j-1) + u(t-1) \tag{8.13}$$

with

$$y(t) = A_Y(t)\beta + H(t)A_X(t-1)\beta$$

$$+ H(t)\sum_{j=1}^{t-1} F(t-1)\cdots F(j)A_X(j-1)\beta$$

$$+ H(t)\sum_{j=1}^{t-1} F(t-1)\cdots F(j)u(j-1)$$

$$+ H(t)u(t-1) + e(t)$$

$$= \mathrm{E}[y(t)]$$

$$+ H(t)\sum_{j=1}^{t-1} F(t-1)\cdots F(j)u(j-1)$$

$$+ H(t)u(t-1) + e(t). \tag{8.14}$$

We may separate the right hand sides in each of expressions (8.12) and (8.14) into two parts: one corresponding to the mean of the left hand side and another term that is identical to the parallel expression in our original, zero mean, state-space model (1.19)–(1.28) having

$S(0|0) = 0$. For example, in (8.12) we see that once the mean is removed from $x(t)$ we are left with

$$x_0(t) = \sum_{j=1}^{t-1} F(t-1) \cdots F(j)u(j-1) + u(t-1)$$

which is recognized, for example, from Section 6.2.1 as the state vector that arises from the zero mean state-space model (6.7)–(6.8) where $S(0|0) = 0$.

The most immediate implication of the above developments is that, modulo the incorporation of suitable mean adjustments, all of our previous algorithms can be used almost without change to compute BLUPs of the state and signal vectors and their prediction error variance-covariance matrices. To see how this can be accomplished let us first consider how to conduct the forward or filtering part of the computations.

From Theorem 1.1 we have

$$x(t|t-1) = E[x(t)] + \sum_{j=1}^{t-1} \text{Cov}(x(t), \varepsilon(j))R^{-1}(j)\varepsilon(j).$$

But, from (8.5) and (8.14) we see that

$$\varepsilon(t) = y(t) - E[y(t)]$$

$$-H(t)\sum_{j=1}^{t-1} \text{Cov}(x(t), \varepsilon(j))R^{-1}(j)\varepsilon(j)$$

$$= y(t) - A_Y(t)\beta - H(t)\Big[E[x(t)]$$

$$+ \sum_{j=1}^{t-1} \text{Cov}(x(t), \varepsilon(j))R^{-1}(j)\varepsilon(j)\Big]$$

$$= y(t) - A_Y(t)\beta - H(t)x(t|t-1)$$

and, from (8.4) and (8.12),

$$x(t|t-1) = A_X(t-1)\beta + F(t-1)\Big[\mathrm{E}[x(t-1)]$$

$$+ \sum_{j=1}^{t-1} \mathrm{Cov}(x(t-1), \varepsilon(j))R^{-1}(j)\varepsilon(j)\Big]$$

$$= A_X(t-1)\beta + F(t-1)x(t-1|t-1).$$

Thus, we obtain a slight twist on our previous recursive filtering formulae by taking

$$\varepsilon(1) = y(1) - A_Y(1)\beta - H(1)A_X(0)\beta,$$

$$x(1|1) = A_X(0)\beta + S(1|0)H^T(1)R^{-1}(1)\varepsilon(1)$$

and, for $t = 2, \ldots, n$,

$$x(t|t-1) = A_X(t-1)\beta + F(t-1)x(t-1|t-1),$$

$$\varepsilon(t) = y(t) - A_Y(t)\beta - H(t)x(t|t-1)$$

with

$$x(t|t) = \mathrm{E}[x(t)] + \sum_{j=1}^{t} \mathrm{Cov}(x(t), \varepsilon(j))R^{-1}(j)\varepsilon(j)$$

$$= A_X(t-1)\beta + \mathrm{Cov}(x(t), \varepsilon(t))R^{-1}(t)\varepsilon(t)$$

$$+ F(t-1)\Big[\mathrm{E}[x(t-1)]$$

$$+ \sum_{j=1}^{t-1} \mathrm{Cov}(x(t-1), \varepsilon(j))R^{-1}(j)\varepsilon(j)\Big]$$

$$= A_X(t-1)\beta$$

$$+ S(t|t-1)H^T(t)R^{-1}(t)\varepsilon(t)$$

$$+ F(t-1)x(t-1|t-1)$$

$$= S(t|t-1)H^T(t)R^{-1}(t)\varepsilon(t) + x(t|t-1).$$

The smoothing step is actually somewhat simpler to derive in this context because when $r > t$

$$x(t|r) \; = \; E[x(t)] + \sum_{j=1}^{t} \mathrm{Cov}(x(t), \varepsilon(j)) R^{-1}(j)\varepsilon(j)$$

$$+ \sum_{j=t+1}^{r} \mathrm{Cov}(x(t), \varepsilon(j)) R^{-1}(j)\varepsilon(j)$$

$$= \; x(t|t) + \sum_{j=t+1}^{r} \mathrm{Cov}(x(t), \varepsilon(j)) R^{-1}(j)\varepsilon(j).$$

The second term in this last expression is exactly the same as what is computed by the fixed interval smoothing Algorithm 5.1 and, accordingly, needs no special treatment beyond what we have done before.

The following algorithm illustrates how the ideas in this section can be developed to produce a combined version of Algorithms 4.3 and 5.1 that is applicable to data from model (8.1)–(8.3).

Algorithm 8.1 This algorithm returns $x(t|t)$, $S(t|t)$, $f(t|t)$, $V(t|t)$, $t = 1, \ldots, n$, and, for specified r, evaluates $x(t|r)$, $S(t|r)$, $f(t|r)$, $V(t|r)$, $t = 1, \ldots, r$.

/*Initialization of forward recursion*/
$R(1) = H(1)Q(0)H^{T}(1) + W(1)$
$S(1|1) = Q(0) - Q(0)H^{T}(1)R^{-1}(1)H(1)Q(0)$
$\varepsilon(1) = y(1) - [A_Y(1) + H(1)A_X(0)]\beta$
$x(1|1) = A_X(0)\beta + Q(0)H^{T}(1)R^{-1}(1)\varepsilon(1)$
for $t = 2$ **to** n
    $S(t|t-1) = F(t-1)S(t-1|t-1)F^{T}(t-1)$
        $+Q(t-1)$
    $R(t) = H(t)S(t|t-1)H^{T}(t) + W(t)$
    $S(t|t) = S(t|t-1)$
        $-S(t|t-1)H^{T}(t)R^{-1}(t)H(t)S(t|t-1)$
    $V(t|t) = H(t)S(t|t)H^{T}(t)$

$$x(t|t-1) = A_X(t-1)\beta$$
$$+F(t-1)x(t-1|t-1)$$
$$\varepsilon(t) = y(t) - A_Y(t)\beta - H(t)x(t|t-1)$$
$$x(t|t) = S(t|t-1)H^T(t)R^{-1}(t)\varepsilon(t)$$
$$+x(t|t-1)$$
$$f(t|t) = H(t)x(t|t)$$

end for
/*Initialization of smoothing recursion*/
$$a = M^T(r-1)H^T(r)R^{-1}(r)\varepsilon(r)$$
$$A = M^T(r-1)H^T(r)R^{-1}(r)H(r)M(r-1)$$

for $t = r-1$ **to** 1
$$x(t|r) = x(t|t) + S(t|t-1)a$$
$$S(t|r) = S(t|t) + S(t|t-1)AS(t|t-1)$$
$$f(t|r) = H(t)x(t|r)$$
$$V(t|r) = H(t)S(t|r)H^T(t)$$
$$a = M^T(t-1)H^T(t)R^{-1}(t)\varepsilon(t)$$
$$+M^T(t-1)a$$
$$A = M^T(t-1)H^T(t)R^{-1}(t)H(t)M(t-1)$$
$$+M^T(t-1)AM(t-1)$$

end for

8.3   Estimation of β

In this section we address the issue of estimating the vector β in the case where it is unknown. We will accomplish this using a least-squares approach much as we did in Section 6.3. Accordingly, we first need to derive expressions for the mean and variance-covariance matrix of the response vector $y = (y^T(1), \ldots, y^T(n))^T$ and the state vector $x = (x^T(1), \ldots, x^T(n))^T$.

We will once again need the matrices H and F defined in (6.5) and (6.10) of Section 6.2.1. Using these two matrices we can write model (8.1)–(8.2) as

$$y = G\beta + HFu + e,$$

$$x = T\beta + Fu,$$

where $u = (u^T(0), \ldots, u^T(n-1))^T$,

$$T = FA_X, \tag{8.15}$$

$$G = A_Y + HT, \tag{8.16}$$

$$A_X = \begin{bmatrix} A_X(0) \\ A_X(1) \\ \vdots \\ A_X(n-1) \end{bmatrix}, \tag{8.17}$$

and

$$A_Y = \begin{bmatrix} A_Y(1) \\ A_Y(2) \\ \vdots \\ A_Y(n) \end{bmatrix}. \tag{8.18}$$

Consequently, the mean of x is $T\beta$, the mean of y is $G\beta$ and the x and y variance-covariance matrices are

$$\mathrm{Var}(x) = FQF^T = \Sigma_{x_0},$$

$$\mathrm{Var}(y) = H\Sigma_{x_0}H^T + W = \Sigma_{y_0}$$

with Q and W defined as in (6.9) and (6.6), respectively. Notice that the matrices Σ_{x_0} and Σ_{y_0} are identical to the ones in (6.11) and (6.12) that we encountered in Section 6.2.1 where we dealt with prediction under model (1.19)–(1.28) with $S(0|0) = 0$. We will see that techniques for carrying out prediction in this special case also play a fundamental role in computing predictions, etc., for the model (8.1)–(8.3).

Since $y - G\beta$ has mean zero we can use Theorem 1.1 to obtain a vector-matrix form for the BLUP of x based on y. Specifically, we have

$$\hat{x}(\beta) = \mathrm{E}[x] + \Sigma_{x_0} H^T \Sigma_{y_0}^{-1} (y - G\beta)$$

$$= T\beta + \Sigma_{x_0} H^T \Sigma_{y_0}^{-1} (y - G\beta) \qquad (8.19)$$

because

$$\mathrm{Cov}(x, y - G\beta) = \mathrm{Cov}(x - T\beta, y - G\beta) = \Sigma_{x_0} H^T.$$

We have used the notation $\hat{x}(\beta)$ here to explicitly indicate that this estimator cannot be computed without the specification of some value for β.

Expression (8.19) reveals an alternative, equivalent way to produce the BLUP of x based on y. First, one applies the fixed interval smoothing Algorithm 5.1 with $r = n$ and $S(0|0) = 0$ to y and each column of the matrix G to produce

$$\hat{x}_0 = \Sigma_{x_0} H^T \Sigma_{y_0}^{-1} y$$

and

$$\hat{G}_0 = \Sigma_{x_0} H^T \Sigma_{y_0}^{-1} G.$$

Then, $\hat{x}(\beta)$ can be obtained from

$$\hat{x}(\beta) = \hat{x}_0 + (T - \hat{G}_0)\beta. \qquad (8.20)$$

The algorithm that results from this will require only order n flops provided that G and T are evaluated efficiently. This can be accomplished via the relations

$$T(t) = A_X(t-1)$$

$$+ \sum_{j=1}^{t-1} F(t-1) \cdots F(j) A_X(j-1)$$

$$= A_X(t-1) + F(t-1)T(t-1) \qquad (8.21)$$

and

$$G(t) = A_Y(t) + H(t)T(t) \qquad (8.22)$$

with initialization via $T(1) = A_X(0)$.

Now consider the problems of predicting x in the case where β is unknown. Following the strategy we employed in Section 6.3, the BLUE of β is

$$\widehat{\beta} = (G^T \operatorname{Var}^{-1}(y)G)^{-1} G^T \operatorname{Var}^{-1}(y)y$$
$$= (G^T \Sigma_{y_0}^{-1} G)^{-1} G^T \Sigma_{y_0}^{-1} y \qquad (8.23)$$

with G now defined in (8.16). In combination with (8.20) this suggests predicting x with

$$\widehat{x} = \widehat{x}(\widehat{\beta})$$
$$= \widehat{x}_0 + (T - \widehat{G}_0)\widehat{\beta}$$
$$= T(G^T \Sigma_{y_0}^{-1} G)^{-1} G^T \Sigma_{y_0}^{-1} y$$
$$\quad + \Sigma_{x_0} H^T \Sigma_{y_0}^{-1} \Big[I$$
$$\quad - G(G^T \Sigma_{y_0}^{-1} G)^{-1} G^T \Sigma_{y_0}^{-1} \Big] y \qquad (8.24)$$

which is exactly the same as the formula we obtained for the diffuse predictor x_∞ in (6.21) of Section 6.2.2 apart from the fact that the T and G matrices are now defined differently. As a result of this, one may conclude that Theorem 6.2 is equally applicable to this setting with the implication that $\widehat{x}(\widehat{\beta})$ is the BLUP of x with prediction error variance-covariance matrix S having the same form as S_∞ in (6.22) but with T and G defined as in (8.15) and (8.16), respectively. The related predictor of the signal $f = Hx$ is then $\widehat{f}(\widehat{\beta}) = H\widehat{x}(\widehat{\beta})$ with prediction error variance-covariance matrix $V = HSH^T$. To compute $\widehat{x}(\widehat{\beta})$, $\widehat{f}(\widehat{\beta})$ and the diagonal blocks of S, V we can apply Algorithm 6.1 to the y vector from model (8.1)–(8.3) and the columns of G with G being computed recursively from (8.21) and (8.22).

8.4 Likelihood evaluation

To conclude, let us discuss evaluation of the sample likelihood for the case where (8.1)–(8.3) corresponds to normal response and state processes. From our work in the previous section we already know that y has mean vector $G\beta$, for G in (8.16) and variance-covariance Σ_{y_0}. As a result, two times the negative log-likelihood is seen to be

$$\ell = pn \ln 2\pi + \ln |\Sigma_{y_0}|$$
$$+ (y - G\beta)^T \Sigma_{y_0}^{-1} (y - G\beta). \qquad (8.25)$$

The sample log-likelihood can be minimized to obtain an mle for the parameter vector β with the result being precisely the BLUE given by $\hat{\beta}$ in (8.23). Upon replacing β by $\hat{\beta}$ in (8.25) we obtain a concentrated log-likelihood function of the form

$$\ell = pn \ln 2\pi + \ln |\Sigma_{y_0}|$$
$$+ (y - G\hat{\beta})^T \Sigma_{y_0}^{-1} (y - G\hat{\beta})$$
$$= pn \ln 2\pi + \ln |\Sigma_{y_0}| + y^T \Sigma_{y_0}^{-1} y$$
$$- y^T \Sigma_{y_0}^{-1} G (G^T \Sigma_{y_0}^{-1} G)^{-1} G^T \Sigma_{y_0}^{-1} y$$
$$= pn \ln 2\pi$$
$$+ \sum_{t=1}^{n} \left[\ln |R_0(t)| + \varepsilon_0^T(t) R_0^{-1}(t) \varepsilon_0(t) \right]$$
$$- \varepsilon_0^T R_0^{-1} E_0 (E_0^T R_0^{-1} E_0)^{-1} E_0^T R_0^{-1} \varepsilon_0$$
$$(8.26)$$

with $\varepsilon_0 = L_0^{-1} y$ and $E_0 = L_0^{-1} G$ for R_0, L_0 from the

Cholesky factorization $\Sigma_{y_0} = L_0 R_0 L_0^T$. Apart from the difference in the definition of the matrix G, this is exactly the same as the likelihood function we dealt with at the end of Section 7.3. Thus, it can be evaluated efficiently by following the steps laid out in Section 7.2 subsequent to (7.3). Further concentration of the likelihood with respect to estimation of a scale parameter results in an estimator of the same form as (7.6) in Section 7.4. Finally, as we demonstrated in the simpler setting of Chapter 7, it is also the case for model (8.1)–(8.3) that a diffuse specification for β produces a sample likelihood that agrees with (8.26) apart from the addition of the factor $\ln|G^T \Sigma_{y_0}^{-1} G| = \ln|E_0^T R_0^{-1} E_0|$. A detailed proof of this can be found in de Jong (1988).

A

The Cholesky Decomposition

Given the central role that Cholesky factorization plays
in Kalman filtering, this book would not be complete
without a detailed discussion of the Cholesky method.
This is provided by the material that follows.

To begin, let $\Sigma = \{\sigma(i,j)\}$ be a $np \times np$, positive definite matrix with $p \times p$ sub-blocks $\sigma(i,j)$, $i,j = 1, \ldots, n$.
Then, we want to obtain recursive formulae for evaluat-
ing the matrices L and R in the decomposition

$$\Sigma = LRL^T$$

with $R = \mathrm{diag}(R(1), \ldots, R(n))$ and $L = \{L(i,j)\}$ a
block lower triangular matrix having $L(i,i) = I$, $i =
1, \ldots, n$, for I the $p \times p$ identity matrix.

We also want to use the factorization algorithm to solve
linear systems such as $\Sigma b = z$, where $z = (z^T(1), \ldots,
z^T(n))^T$ is a specified pn-vector and the vector $b =
(b^T(1), \ldots, b^T(n))^T$ represents the unknown solution.
Note, however, that given L and R we have

$$Lc = z \tag{A.1}$$

with $c = RL^T b$. Thus, $c = (c^T(1), \ldots, c^T(n))^T$ may be
obtained via forward solution of a block lower triangular
system. More precisely, since the diagonal blocks of L
are identity matrices, it follows that $c(1) = z(1)$ and,

for $t = 2, \ldots, n,$

$$c(t) = [z(t) - \sum_{i=1}^{t-1} L(t, i)c(i)]. \tag{A.2}$$

Upon obtaining c by recursion (A.2) we can back-solve the block upper triangular system $L^T b = R^{-1} c$ to obtain b. This produces $b(n) = R^{-1}(n)c(n)$ and

$$b(t) = R^{-1}(t)c(t) - \sum_{i=t+1}^{n} L^T(i, t)b(i) \tag{A.3}$$

for $t = n - 1, \ldots, 1.$

To use (A.2)–(A.3) we must first evaluate L and R. For this purpose it suffices to equate the elements of LRL^T and Σ. If we now let $L(1, 1 : n) = [I, 0, \ldots, 0]$ be the first row block of L and, similarly, define

$$L(t, 1 : n) = [L(t, 1), \ldots, L(t, t - 1), I, 0, \ldots, 0],$$

the problem becomes tantamount to finding choices for $L(1, 1 : n), \ldots, L(n, 1 : n)$ and R that satisfy

$$L(t, 1 : n)RL^T(j, 1 : n) = \sigma(t, j), \tag{A.4}$$

for $j = 1, \ldots, n$ and $t \geq j$ since Σ is symmetric.

Starting with the upper left hand corner of Σ and working downwards (i.e., with $j = 1$ and $t \geq 1$ in (A.4)) we obtain

$$R(1) = \sigma(1, 1), \tag{A.5}$$

$$L(t, 1) = \sigma(t, 1)R^{-1}(1), \tag{A.6}$$

for $t = 2, \ldots, n$ and, in general, have

$$R(j) = \sigma(j, j) - \sum_{i=1}^{j-1} L(j, i)R(i)L^T(j, i) \tag{A.7}$$

with

$$L(t, j) = [\sigma(t, j) - \sum_{i=1}^{j-1} L(t, i)R(i)L^T(j, i)]R^{-1}(j)$$

(A.8)

for $t = j + 1, \ldots, n$. Since all the solutions provided by (A.5)–(A.8) are determined uniquely it follows that the Cholesky decomposition is itself unique.

Observe that each step in (A.5)–(A.8) involves only those sub-matrices that were computed on the previous iteration. Also, note the similarity between (A.2) and (A.7) which has the practical implication that the forward solution step can be done in tandem with the computation of L and R.

If we now think a bit more about (A.1)–(A.2) we can see that the forward solution step actually produces $c = L^{-1}z$ for any "input" vector z. In this respect there is nothing special about having a single vector for the right hand side and more generally we can take $Z = [Z^T(1, 1: r), \ldots, Z^T(n, 1: r)]^T$ for $p \times r$ matrices $Z(t, 1: r)$, $t = 1, \ldots, n$. Then, to solve $LC = Z$ for the $np \times r$ matrix C we can use $C(1, 1: r) = Z(1, 1: r)$ and, for $t = 2, \ldots, n$,

$$C(t, 1: r) = \left[Z(t, 1: r) - \sum_{i=1}^{t-1} L(t, i)C(i, 1: r) \right]. \quad (A.9)$$

The back substitution step proceeds in a similar manner with $B(n, 1: r) = R^{-1}(n)C(n, 1: r)$ and

$$B(t, 1: r) = R^{-1}(t)C(t, 1: r) - \sum_{i=t+1}^{n} L^T(i, t)B(i, 1: r).$$

(A.10)

A complete description of the Cholesky method that allows for multiple right hands sides is given below.

Algorithm A.1 This algorithm computes the diagonal blocks of R and the below diagonal blocks of L in the

Cholesky factorization $\Sigma = LRL^T$ and also returns the solution to the linear system $\Sigma B = Z$.

```
/*Initialization*/
R(1) = σ(1, 1)
C(1, 1 : r) = Z(1, 1 : r)
/*Computation of first column*/
for t = 2 to n
    L(t, 1) = σ(t, 1)R⁻¹(1)
end for
/*Computation of remaining columns*/
for j = 2 to n
    /*Compute R(j) first*/
    R(j) = σ(j, j)
    C(j, 1 : r) = Z(j, 1 : r)
    for i = 1 to j − 1
        R(j) := R(j) − L(j, i)R(i)Lᵀ(j, i)
        C(j, 1 : r) := C(j, 1 : r) − L(j, i)C(i, 1 : r)
    end for
    /*Now compute the rest of the column*/
    for t = j + 1 to n
        L(t, j) = σ(t, j)
            for i = 1 to j − 1
                L(t, j) := L(t, j)
                          −L(t, i)R(i)Lᵀ(j, i)
            end for
        L(t, j) := L(t, j)R⁻¹(j)
    end for
end for
/*Initialization for back substitution*/
B(n, 1 : r) = R⁻¹(n)C(n, 1 : r)
/*Back Substitution*/
for t = n − 1 to 1
    B(t, 1 : r) = R⁻¹(t)C(t, 1 : r)
    for j = t + 1 to n
        B(t, 1 : r) := B(t, 1 : r) − Lᵀ(j, t)B(j, 1 : r)
    end for
end for
```

In this book we are most interested in situations where n is much larger than p which makes it the dominant factor for determining computation times. In this respect, if we notationally suppress the influence of p, the number of flops (floating point operations) required in the jth step of the Cholesky recursion (A.5)–(A.8) is seen to be on the order of $j(n - j)$. Since $\sum_{j=1}^{n} j(n - j) = n(n^2 - 1)/6$, the total number of flops is $O(n^3)$. The same remains true for solving linear systems using the Cholesky method.

As a special case of (A.9) we can choose Z to be an $np \times p$ matrix of all zeros except for $Z(i)$ which we take to be the p-dimensional identity. Our matrix forward recursion, omitting multiplication by R^{-1}, will then produce the $p \times p$ block matrices that correspond to the ith (block) column of $L^{-1} = \{L^{-1}(t, j)\}_{t,j=1:n}$. A rote application of this idea produces the formulae

$$L^{-1}(t, j) = -\sum_{i=1}^{t-1} L(t, i)L^{-1}(i, j)$$

for $j \neq t$ and

$$L^{-1}(j, j) = I - \sum_{i=1}^{j-1} L(j, i)L^{-1}(i, j).$$

However, there are simplification here because L^{-1} must also be lower triangular with identity matrix diagonal blocks. Through use of these properties we can establish the following result.

Theorem A.1 *Let $L^{-1} = \{L^{-1}(t, j)\}_{t,j=1:n}$ for $p \times p$ block matrices $L^{-1}(t, j), t, j = 1, \ldots, n$. Then, $L^{-1}(t, t) = I$ for $t = 1, \ldots, n$, $L^{-1}(t, j) = 0$ for $j > t$,*

$$L^{-1}(t, t - 1) = -L(t, t - 1), \qquad \text{(A.11)}$$

for $t = 2, \ldots, n$ *and*

$$L^{-1}(t, j) = -L(t, j) - \sum_{i=j+1}^{t-1} L(t, i)L^{-1}(i, j) \quad (A.12)$$

for $t = 3, \ldots, n, \; j = 1, \ldots, t - 2.$

Proof. It is worthwhile to go through a bit more detail on establishing this result beyond just a reference to (A.9). The idea is that we are solving the system $LB = I$ for the matrix B to obtain the solution $B = L^{-1}$. Since L is lower triangular with identity matrix diagonal blocks, this immediately fixes $L^{-1}(1, 1)$ as the identity and all the remaining row blocks as zero matrices. One may now proceed by induction to see that the diagonal blocks of L^{-1} must all be identities and that its above diagonal blocks are all zero matrices.

At this point a more enlightened application of (A.9) (using $L^{-1}(t, j) = 0$ for $j > t$) reveals that for $t > j$

$$L^{-1}(t, j) = - \sum_{i=j}^{t-1} L(t, i)L^{-1}(j, i)$$

because $L(t, t) = I$. Using $L^{-1}(j, j) = I$ completes the proof. ∎

As a result of Theorem A.1 we see that the the evaluation of the below diagonal row blocks for L^{-1} can proceed by first computing $L^{-1}(2, 1) = -L(2, 1)$ and then $L^{-1}(3, 2) = -L(3, 2)$ which allows for evaluation of

$$L^{-1}(3, 1) = -L(3, 1) - L(3, 2)L^{-1}(2, 1).$$

Next we compute $L^{-1}(4, 3) = -L(4, 3)$ which leads to

$$L^{-1}(4, 2) = -L(4, 2) - L(4, 3)L^{-1}(3, 2)$$

and

$$L^{-1}(4, 1)$$

$$= -L(4, 1) - L(4, 2)L^{-1}(2, 1) - L(4, 3)L^{-1}(3, 1).$$

By proceeding in this fashion we can formulate the following computational scheme.

Algorithm A.2 This algorithm computes the below diagonal blocks of L^{-1}.

```
for t = 2 to n
    for j = 1 to t − 1
        L⁻¹(t, j) = −L(t, j)
        if j < t − 1
            for i = j + 1 to t − 1
                L⁻¹(t, j) := L⁻¹(t, j)
                              −L(t, i)L⁻¹(i, j)
            end for
        end if
    end for
end for
```

B

Notation Guide

0	a matrix of all zero entries
I	the identity matrix
$A(i, j)$	the block of the matrix A corresponding to its ith block row and jth block column
E	mathematical expectation
$\text{Var}(z)$	variance-covariance matrix for z
$\text{Cov}(z, v)$	covariance matrix for z and v
n	the number of response vectors
t	an integer valued "time" index
p	the dimension of each response vector
$y(t)$	observed response vector at the time index t: pg. 3
y	the stacked vector of all the responses: pg. 3
$f(t)$	unobserved $p \times 1$ signal vector at time t
f	the stacked vector of all the signals: pg. 11
$e(t)$	unobserved noise vector at time t

$W(t)$	$e(t)$'s variance-covariance matrix
e	the stacked vector of all the noise variables: pg. 11
W	variance-covariance matrix for e: pg. 4
q	dimension of the state vectors
$x(0)$	initial state vector
$S(0|0)$	$x(0)$'s variance-covariance matrix
$x(t)$	the state-vector at time index t
x	the stacked vector of all the state variables: pg. 110
$u(t)$	a $q \times 1$ vector of random disturbances
$Q(t)$	$u(t)$'s variance-covariance matrix
u	the stacked vector of all the disturbances: pg. 126
Q	variance-covariance matrix for u: pg. 113
$\varepsilon(t)$	innovation vector at time index t: pg. 8
$R(t)$	$\varepsilon(t)$'s variance-covariance matrix
ε	the stacked vector of all the innovations: pg. 10
R	variance-covariance matrix for ε: pg. 10
$F(t)$	state transition matrix
$H(t)$	state transformation matrix
$x(t|j)$	least-squares linear predictor of $x(t)$ from $\varepsilon(1), \ldots, \varepsilon(j)$: pg. 13

$S(t|j)$ — prediction error variance-covariance matrix for $x(t|j)$: pg. 14

$f(t|j)$ — the least-squares linear predictor of $f(t)$ from $\varepsilon(1), \ldots, \varepsilon(j)$: pg. 14

$V(t|j)$ — prediction error variance-covariance matrix for $f(t|j)$: pg. 14

$K(t)$ — the Kalman gain matrix: pg. 54

$M(t)$ — matrix defined in (2.17) on pg. 113

$\sigma_{X\varepsilon}(t, j)$ — the $q \times p$ covariance matrix for $x(t)$ and $\varepsilon(j)$: pg. 30

$\Sigma_{X\varepsilon}$ — the $nq \times np$ of the $\sigma_{X\varepsilon}(t, j)$: pg. 30

L — $np \times np$ lower triangular matrix in the Cholesky decomposition of $\text{Var}(y)$: pg. 10

$L(t, j)$ — $p \times p$ matrix in the tth block row and jth block column of L: pg. 10

L^{-1} — $np \times np$ lower triangular inverse matrix corresponding to the matrix L in the Cholesky decomposition for $\text{Var}(y)$: pg. 58

$L^{-1}(t, j)$ — $p \times p$ matrix in the tth block row and jth block column of L^{-1}: pg. 58

H — the $np \times nq$ block diagonal matrix containing all the $H(t)$: pg. 111

F — the $nq \times nq$ block lower triangular matrix in (6.10) on pg. 113

T — $nq \times q$ matrix in (6.3) on pg. 110

G	the matrix HT: pg. 119
Σx_0	variance-covariance matrix for x when $S(0\|0) = 0$: pg. 113
Σy_0	variance-covariance matrix for y when $S(0\|0) = 0$: pg. 113
L_0	block lower triangular matrix in the Cholesky factorization of the matrix Σy_0: pg. 121
R_0	block diagonal matrix from the Cholesky factorization of Σy_0: pg. 121
ε_0	the vector $\varepsilon_0 = L_0^{-1} y$: pg. 121
E_0	the $np \times q$ matrix $E_0 = L_0^{-1} G$: pg. 121
$\hat{x}_0(v)$	state vector prediction formula for when $S(0\|0) = 0$ applied to a given vector v: pg. 115
$\hat{f}_0(v)$	the signal vector prediction formula for when $S(0\|0) = 0$ applied to a given vector v: pg. 115
S_0	prediction error covariance matrix for the BLUP of the state vector when $S(0\|0) = 0$: pg. 114
V_0	prediction error covariance matrix for the BLUP of the signal when $S(0\|0) = 0$: pg. 114
$\hat{x}_\infty$	diffuse state vector predictor: pg. 119
$\hat{f}_\infty$	diffuse signal predictor: pg. 125

S_∞ prediction error covariance matrix for $\hat{x}_\infty$: pg. 119

V_∞ prediction error covariance matrix for $\hat{f}_\infty$: pg. 125

ℓ twice the negative log-likelihood: pg. 139

References

Anderson, B. D. O. and Moore, J. B. (1979). *Optimal Filtering*. New Jersey: Prentice-Hall.

Ansley, C. F. and Kohn, R. (1985). Estimation, filtering and smoothing in state space models with incompletely specified initial conditions. *The Annals of Statistics* **13**, 1286–1316.

Davis, P. (1975). *Interpolation and Approximation*. New York: Dover.

Doob, J. (1953). *Stochastic Processes*. New York: Wiley.

Eubank, R. and Wang, S. (2002). The equivalence between the Cholesky decomposition and the Kalman filter. *American Statistician* **56**, 39–43.

Gelb, A. (1974). *Applied Optimal Estimation*. Boston: MIT Press.

Goldberger, A. (1962). Best linear unbiased prediction in the generalized linear regression model. *Journal of the American Statistical Association* **57**, 369–375.

Hocking, R. (2003). *Methods and Applications of Linear Models: Regression and the Analysis of Variance*. New York: Wiley.

Householder, A. (1964). *The Theory of Matrices in*

Numerical Analysis. New York: Dover.

de Jong, P. (1988). The likelihood for a state space model. *Biometrika* **75**, 165–169.

de Jong, P. (1989). Smoothing and interpolation with the state-space model. *Journal of the American Statistical Association* **84**, 1085–1088.

de Jong, P. (1991). The diffuse Kalman filter. *The Annals of Statistics* **19**, 1073–1083.

Kalman, R. (1960). A new approach to linear filtering and prediciton problems. *Journal of Basic Engineering* **82**, 34–45.

Kalman, R. and Bucy, R. (1961). New results in linear filtering and prediction theory. *Journal of Basic Engineering* **85**, 95–108.

Khinchin, A. (1997). *Continued Fractions*. New York: Dover.

Kohn, R. and Ansley, C. (1989). A fast algorithm for signal extraction, influence and cross-validation in state space models. *Biometrika* **76**, 65–79.

Koopman, S. and Durbin, J. (2001). *Time Series Analysis by State Space Methods*. New York: Oxford University Press.

Robinson, G. (1991). The BLUP is a good thing: the estimation of random effects. *Statist. Science* **6**, 15–51.

Wall, H. (1948). *Analytic Theory of Continued Fractions*. New York: van Nostrand.

Index

Printed in the United States
by Baker & Taylor Publisher Services

Printed in the United States
by Baker & Taylor Publisher Services